The Pocket Guide to

Fungal Infection

To families and friends

The Pocket Guide to

Fungal Infection

Malcolm D. Richardson
Mycology Unit
Department of Bacteriology & Immunology
Haartman Institute, University of Helsinki
00014 Helsinki, Finland

Elizabeth M. Johnson
Mycology Reference Laboratory
Public Health Laboratory Service
Bristol BS2 8EL, United Kingdom

Blackwell Science

Editorial Offices:
Osney Mead, Oxford OX2 0EL
25 John Street, London WC1N 2BL
23 Ainslie Place, Edinburgh EH3 6AJ
350 Main Street, Malden
MA 02148 5018, USA
54 University Street, Carlton
Victoria 3053, Australia
10, rue Casimir Delavigne
75006 Paris, France

Other Editorial Offices:
Blackwell Wissenschafts-Verlag GmbH
Kurfürstendamm 57
10707 Berlin, Germany

Blackwell Science KK
MG Kodenmacho Building
7–10 Kodenmacho Nihombashi
Chuo-ku, Tokyo 104, Japan

First published 2000

Set by Sparks Computer Solutions Ltd, Oxford, UK
Printed and bound in Italy by Rotolito Lombarda SpA, Milan

DISTRIBUTORS

Marston Book Services Ltd
PO Box 269
Abingdon, Oxon OX14 4YN
(*Orders*: Tel: 01235 465500
Fax: 01235 465555)

USA
Blackwell Science, Inc
Commerce Place
350 Main Street
Malden, MA 02148 5018
(*Orders*: Tel: 800 759 6102
781 388 8250
Fax: 781 388 8255)

Canada
Login Brothers Book Company
324 Saulteaux Crescent
Winnipeg, Manitoba R3J 3T2
(*Orders*: Tel: 204 837-2987)

Australia
Blackwell Science Pty Ltd
54 University Street
Carlton, Victoria 3053
(*Orders*: Tel: 3 9347 0300
Fax: 3 9347 5001)

A catalogue record for this title is available from the British Library and the Library of Congress

ISBN 0-632-05325-9

For further information on Blackwell Science, visit our website:
www.blackwell-science.com

Contents

Introduction

The main aim of this Pocket Guide is to summarize the major features of fungal infections of humans and to provide visual information for each pathogen and the infections they cause in a convenient and practical format.

In this guide we have provided a succinct account of the clinical manifestations, laboratory diagnosis and management of fungal infections found in European, American and Australasian practice. The guide covers problems encountered both in hospitals and general practice and is designed to facilitate rapid information retrieval with representative colour illustrations.

Our reading list of established literature has been carefully selected to permit efficient access to specific aspects of fungal infection. The list of World Wide Web sites allows access to an even greater wealth of information and illustrative material.

Acknowledgements

We would like to thank David W. Warnock and Colin K. Campbell for kindly agreeing that the slides from the *Slide Atlas of Fungal Infection* could be used in this publication. Many of the illustrations were originally derived from collections of slides held at the Mycology Reference Laboratories in Bristol and Glasgow and we wish to thank friends and colleagues for their generosity in making this resource available to us.

Tinea capitis

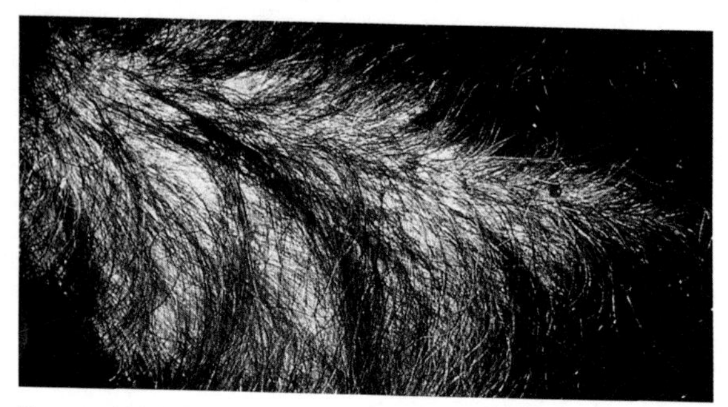

Tinea capitis due to *Trichophyton tonsurans*.

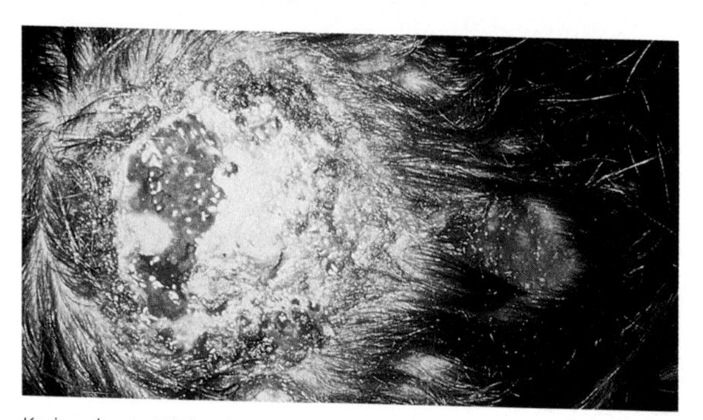

Kerion due to *Trichophyton verrucosum*.

Definition

Tinea capitis describes infection of the scalp and hair with a dermatophyte.

Geographical distribution

World-wide, but more common in Africa, Asia and southern and eastern Europe.

Causal organisms and habitat

- Several *Trichophyton* spp. and *Microsporum* spp.
- Zoophilic *M. canis* is common in western Europe.

- Anthropophilic *T. violaceum* is predominant in eastern and southern Europe and north Africa.
- Anthropophilic *T. tonsurans* is increasing in prevalence.
- Anthropophilic species can be contagious and endemic.
- *T. schoenleinii* causes favus.

Clinical manifestations

- Mild scaling lesions to widespread alopecia.
- Kerion – highly inflammatory, suppurating lesion caused by zoophilic dermatophytes.
- Black dot appearance seen with ectothrix hair invasion.
- Favus is a distinctive infection with grey, crusting lesions.

Essential investigations

Microscopy

Direct microscopic examination of hair roots and skin softened with KOH reveals hyphae, arthrospores and distinctive patterns of hair invasion: ectothrix – large or small arthrospores form a sheath around the hair shaft; endothrix – large or small arthrospores form within the hair shaft; ectoendothrix – spores form around and within the hair shaft; and favus – hyphae and air spaces form within the hairs.

Fluorescence under Wood's light may reveal infected hairs.

Culture

Culture at 28°C for at least 1 week is essential to identify the organism.

Management

Mycological confirmation is essential before commencing oral treatment. Treat with these alternatives:

- griseofulvin 10 mg/kg for up to 3 months
- itraconazole 100 mg/day for 4–6 weeks in adults
- itraconazole pulse therapy for children 5 mg/kg/day for 1 week per month for 3–4 months
- terbinafine 250 mg/day for 4–6 weeks, or longer if *Microsporum canis* is present.

Topical treatment of lesions with an azole may reduce spread.

Tinea corporis

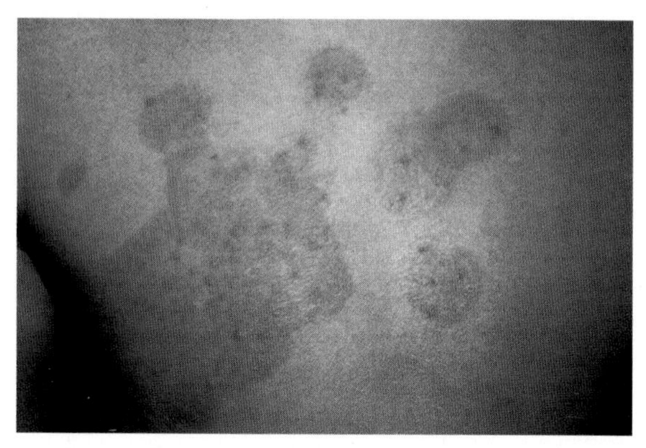

Tinea corporis due to *Trichophyton mentagrophytes*.

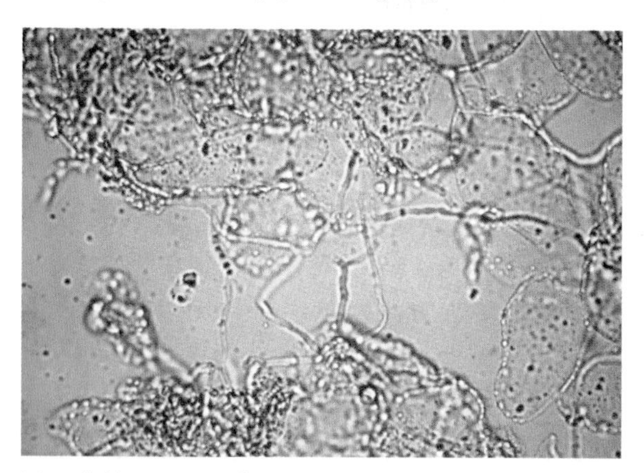

Infected skin scrapings softened in KOH.

Definition

Infection of the skin of the trunk, legs and arms with a dermatophyte.

Geographical distribution

World-wide, but it is more prevalent in tropical and sub-tropical regions.

Causal organisms and habitat

- Many *Trichophyton* spp., *Microsporum* spp. and *Epidermophyton floccosum.*
- Often zoophilic, occasionally geophilic organisms.
- Infection frequently contracted from a household pet.
- May follow infection of another body site.
- *M. canis* from cats and dogs most frequent.
- *T. verrucosum* from cattle in rural areas.

Clinical manifestations

- Usually affects exposed body sites.
- Exact nature depends on infecting organism; infections due to zoophilic species are often more inflammatory and may be pustular.
- Typically, there are itching, dry, circular, scaling lesions.
- Fungus more active at margin therefore more erythematous.

Essential investigations

Microscopy

Skin scrapings should be collected from the raised border. Direct microscopy of skin scrapings softened with KOH reveals branching hyphae with or without arthrospores.

Adhesive tape strippings may be used if little material can be scraped.

Culture

Isolation of the dermatophyte at 28°C allows identification.

Management

This condition seldom resolves if untreated. However, it often responds to topical treatment with an azole (clotrimazole, econazole, miconazole, sulconazole), naftifine or terbinafine morning and evening for 2–4 weeks.

Oral therapy is indicated if the lesions are extensive or refractory. Treat with these alternatives:

- itraconazole 200 mg/day for 1 week
- terbinafine 250 mg/day for 2–4 weeks
- griseofulvin 10 mg/kg for 4 weeks.

Tinea cruris

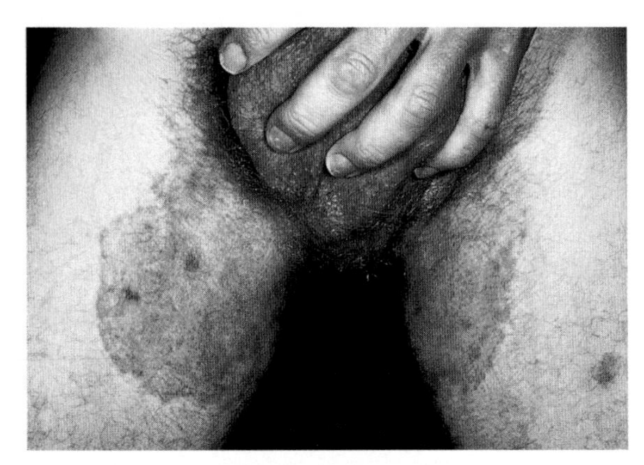

Tinea cruris due to *Trichophyton rubrum*.

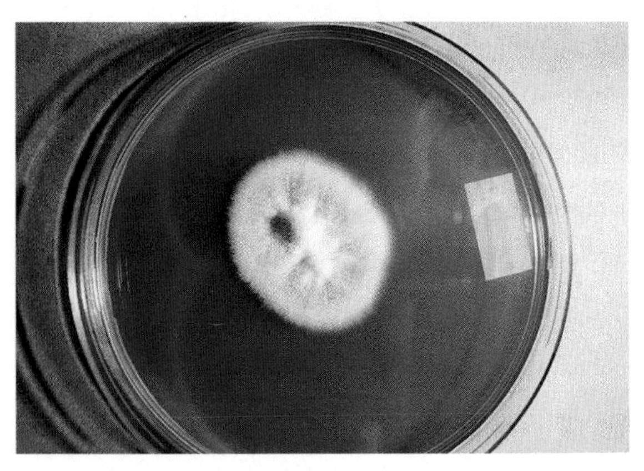

Colonial appearance of *Trichophyton rubrum*.

Definition

Infection of the skin of the groin and pubic region with a dermatophyte.

Geographical distribution

World-wide.

Causal organisms and habitat

- Anthropophilic dermatophytes *Epidermophyton floccosum* and *Trichophyton rubrum* are most common.
- Maceration and occlusion of groin skin gives rise to infection.
- Often transferred from another infected body site.
- Highly contagious via contaminated towels, floors, etc.

Clinical manifestations

- One or more rapidly spreading erythematous lesions with central clearing on the inside of the thighs, intense pruritis.
- Lesions with raised erythematous border and brown scaling.
- Infection may extend locally and spread to other body sites.

Essential investigations

Microscopy

Direct microscopy of skin scrapings softened with KOH reveals branching hyphae with or without arthrospores.

Culture

Isolation of the dermatophyte at 28°C allows identification.

Management

This condition seldom resolves if untreated. However, it often responds to topical treatment with an azole (clotrimazole, econazole, miconazole, sulconazole), naftifine or terbinafine morning and evening for 2–4 weeks.

Oral therapy, if indicated, includes these alternatives:

- itraconazole 200 mg/day for 1 week
- terbinafine 250 mg/day for 2–4 weeks
- griseofulvin 10 mg/kg for 4 weeks.

Hygiene measures such as thorough drying and using separate towels for the groin area should prevent spread.

There is a recurrence in 20–25% of patients.

Tinea pedis

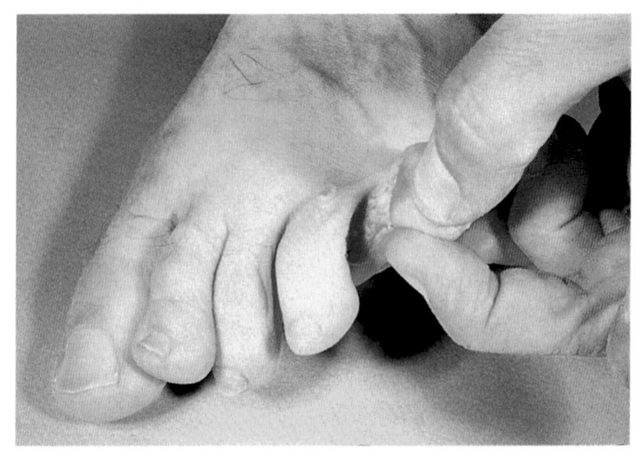

Interdigital tinea pedis due to *Trichophyton rubrum*.

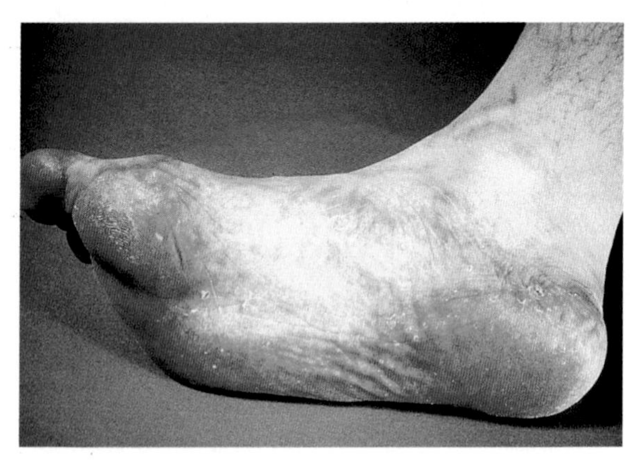

Moccasin form of tinea pedis.

Definition

Dermatophyte infection of the feet.

Geographical distribution

World-wide but more common in countries where there is ready access to communal sports or bathing facilities.

Causal organisms and habitat

- *Trichophyton rubrum* is the most common cause.
- *Epidermophyton floccosum* and *T. mentagrophytes* var. *interdigitale* are also seen.
- Common condition often contracted by walking barefoot on contaminated floors.
- Extensive sweating and occlusive footwear predispose to the condition.
- Infection with the moulds *Scytalidium dimidiatum* (*Hendersonula toruloidea*) and *S. hyalinum* is clinically indistinguishable.

Clinical manifestations

Three types are recognized:

- acute or chronic interdigital infection – itching, peeling, maceration and fissuring of toe webs
- chronic hyperkeratotic (moccasin or dry type) – fine, white scaling limited to heels, soles and lateral borders of feet
- vesicular (inflammatory) infection – vesicle formation on soles, instep and interdigital cleft.

Secondary bacterial or yeast infection is also possible.

Essential investigations

Microscopy

Direct microscopy of skin scrapings softened with KOH reveals branching hyphae with or without arthrospores.

Culture

Isolation of the dermatophyte at 28°C allows identification.

Management

This condition seldom resolves if untreated. However, it often responds to topical treatment with an azole (clotrimazole, econazole, miconazole, sulconazole), naftifine or terbinafine morning and evening for 2–4 weeks.

Oral therapy, if indicated, includes these alternatives:

- itraconazole 200–400 mg/day for 1 week
- terbinafine 250 mg/day for 2–6 weeks.

Tinea manuum

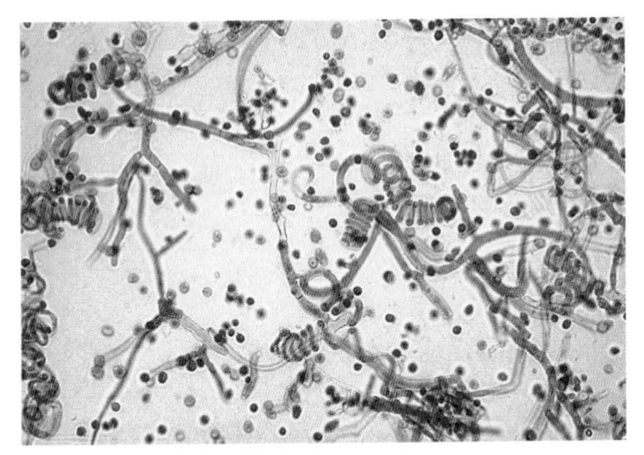

Microscopic appearance of *Trichophyton mentagrophytes*.

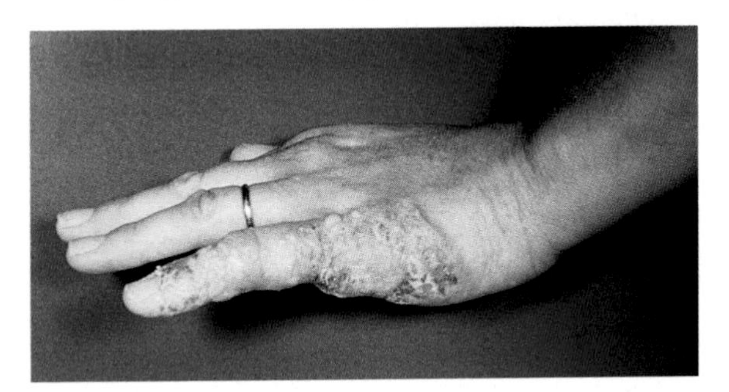

Tinea manuum due to *Trichophyton erinacea*.

Definition

Fungal infection of the hand, or hands.

Geographical distribution

World-wide.

Causal organisms and habitat

- Most common anthropophilic dermatophytes are

Trichophyton mentagrophytes var. *interdigitale*, *T. rubrum* and *Epidermophyton floccosum*.

• Most common zoophilic dermatophytes are *Microsporum canis*, *T. verrucosum* and *T. mentagrophytes* var. *mentagrophytes*.

• Occasional infections due to geophilic *M. gypseum* and *M. fulvum*.

• Acquisition by contact with infected person, animal, soil or fomites, or by autoinoculation from another infected body site.

• Profuse sweating and eczema predispose to infection.

Clinical manifestations

- Usually unilateral, predominantly affecting right hand.
- Two forms: dyshidrotic (eczematoid) and hyperkeratotic:
  - dyshidrotic: annular or segmental vesicles with scaling borders containing clear, viscous fluid on palms, palmar aspect of fingers and sides of the hand, characterized by intense pruritis and burning
  - hyperkeratotic: adjacent vesicles desquamate to form an erythematous, scaling lesion with a circular or irregular thick, white, squamous margin with extensions towards the centre. Chronic cases may cover the entire palm and fingers with fissuring in the palmar creases.

Essential investigations

Microscopy

Direct microscopic examination of vesicle tops and skin scales.

Culture

Isolation in culture at 28°C for at least 1 week will permit species identification.

Management

Topical treatment with imidazole or allylamine is often effective:

- itraconazole 200–400 mg/day for 1 week
- oral terbinafine 250 mg/day for 2–6 weeks.

Tinea unguium

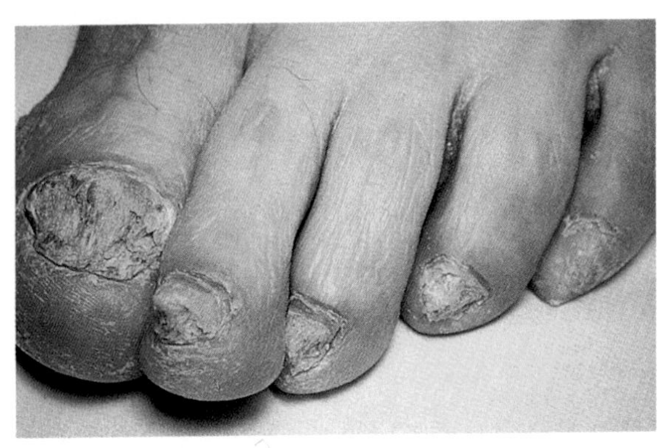

Tinea unguium due to *Trichophyton rubrum*.

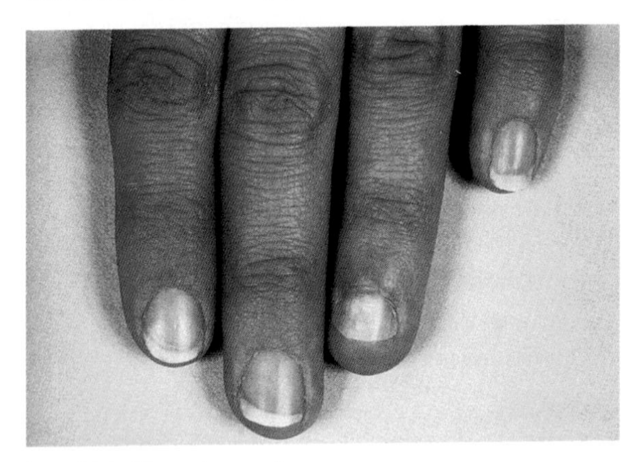

Superficial white onchomycosis.

Definition

Dermatophyte infection of the fingernails or toenails. Onychomycosis is also used to describe the condition but has a broader definition encompassing nail infections with yeasts and moulds in addition to dermatophytes.

Geographical distribution

World-wide.

Causal organisms and habitat

- Most commonly caused by anthropophilic species *T. mentagrophytes* var. *interdigitale* and *T. rubrum*.
- May be rare infections of fingernails with zoophilic species.
- Affects up to 8% of adult population.

Clinical manifestations

- Toenails more often infected than fingernails.
- Infection often follows infection of another body site.
- First and fifth toenails most commonly infected probably due to traumatic damage by ill-fitting footwear.
- White or yellow irregular lesion appears first at free end of nail and spreads slowly to cause entire nail to become thickened, opaque and yellow in colour and may crumble.
- Superficial white onychomycosis is seen predominantly in patients with AIDS where crumbling white lesions, most often due to *T. mentagrophytes* var. *interdigitale,* appear on the nail surface.

Essential investigations

Microscopy

Microscopy of material from KOH softened nails is essential.

Culture

Culture at 28°C will allow identification of the infecting species.

Culture of material on plates with and without cycloheximide will allow differentiation of dermatophyte and mould infections.

Subungual material may be most productive and a nail drill or scalpel may be used.

Management

This condition is difficult to treat, requiring prolonged courses.

- Topical: amorolfine at weekly intervals or tioconazole twice daily for 6 months for fingernails and 9–12 months for toenails.
- Oral: itraconazole 2 or 3 pulse treatment 400 mg/day for 1 week in 4, or continuous 200 mg/day for 3 months; terbinafine 250 mg/day for 6 weeks to 3 months; griseofulvin 4–8 months but low success rate in toenail infections.

Oral candidosis

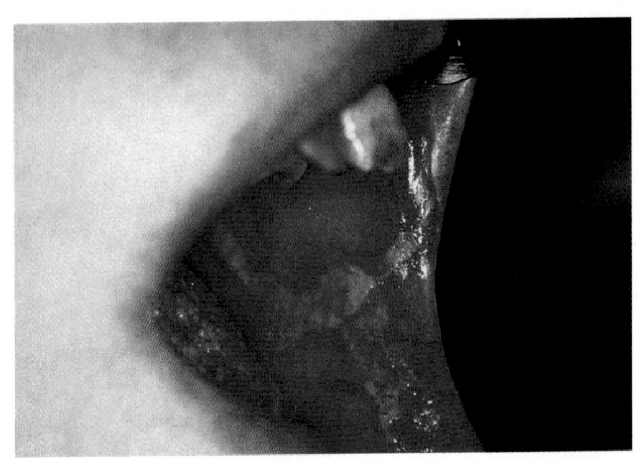

Pseudomembranous oral candidosis.

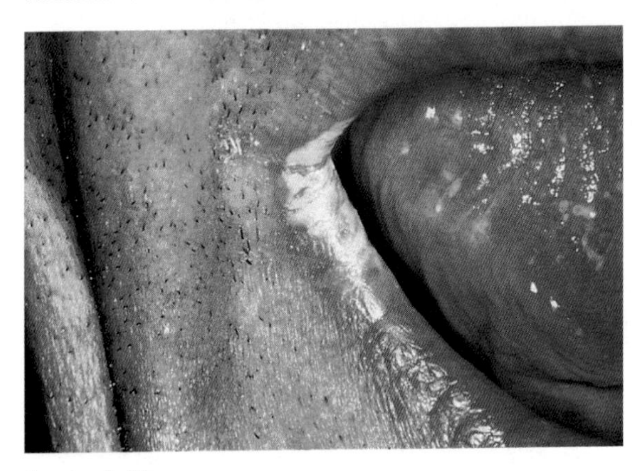

Angular chelitis.

Definition

Oral candidosis is an opportunistic infection of the oral cavity with yeasts of the genus *Candida*. It can be classified into a number of distinct clinical forms.

Geographical distribution

World-wide.

Causal organisms and habitat

- *Candida albicans* is most frequent cause (60–80% cases).
- Eight other pathogenic species.
- Opportunistic infection, often endogenous in origin.
- Infection follows humoral or cell-mediated immunological impairment, debilitation or occlusion of oral surfaces.

Clinical manifestations

- Acute pseudomembranous candidosis (thrush): white raised lesions on buccal mucosa, gums or tongue (infants, the elderly, HIV-infected, diabetics, cancer patients and steroid users).
- Erythematous candidosis: erythema, oedema
  - acute atrophic candidosis – antibiotic stomatitis
  - chronic atrophic candidosis – denture stomatitis and glossitis.
- Chronic hyperplastic candidosis (oral leukoplakia): possibly premalignant, translucent or dense, opaque white plaques on cheeks or tongue associated with smoking.
- Angular chelitis: soreness, erythema and fissuring at corners of mouth often associated with ill-fitting dentures.
- Chronic mucocutaneous candidosis: oral manifestation of candidosis in patients with congenital immunological or endocrinological disorders (see p. 19).

Essential investigations

Microscopy and culture

Microscopy of oral smear reveals filaments and/or yeasts. Culture at 37°C yields the organism in 24–48 hours.

Management

Treat uncomplicated oral candidosis with topical nystatin, amphotericin B or azole for 2 weeks and oral hygiene measures.

Treat immunocompromised patients with these alternatives:

- fluconazole 100–200 mg/day for 2 weeks or 400–800 mg/day for recalcitrant infections
- itraconazole 200–400 mg/day for 1–2 weeks
- parenteral amphotericin B 0.5–0.7 mg/kg/day for 1 week.

Vaginal candidosis

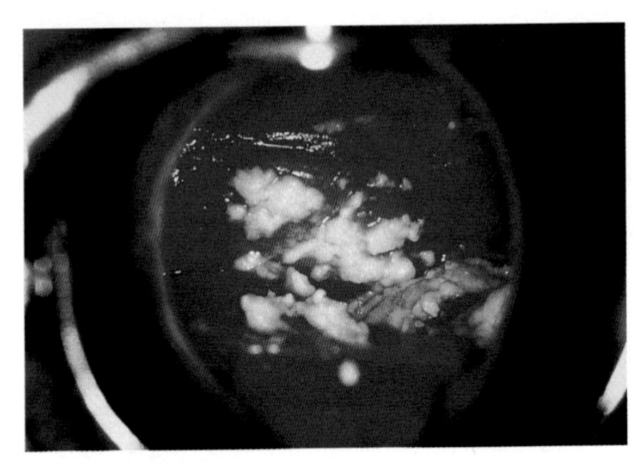

White plaques on vaginal membrane.

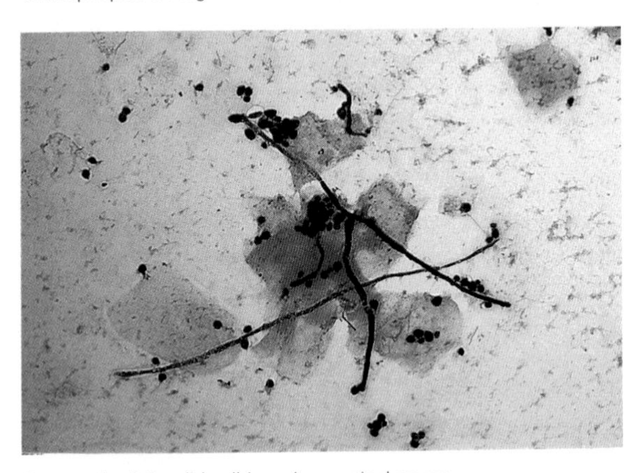

Gram stained *Candida albicans* in a vaginal smear.

Definition

Infection of the vagina and/or vulva with species belonging to the genus *Candida*. Often referred to as vaginal 'thrush'.

Geographical distribution

World-wide.

Causal organisms and habitat

- *Candida albicans* accounts for more than 80% of infections.
- Infections with *C. glabrata* are often refractory to treatment.
- *Candida* spp. can be present in the absence of disease.
- Pregnancy may precipitate chronic or recurrent infections.
- More common amongst women with diabetes mellitus.
- Antibiotic treatment predisposes to yeast overgrowth.
- Treatment of male partner does not seem to prevent recurrence.

Clinical manifestations

- Intense vulval and vaginal pruritis and burning with thick, white, curdy vaginal discharge and adherent white plaques on vulval, vaginal or cervical epithelium.
- Dysuria and dyspareunia are common.
- Perianal intertrigo with pustular or vesicular lesions may occur.
- Tends to occur in the week prior to menstruation.

Essential investigations

Microscopy

Microscopy of a high vaginal swab reveals yeasts with or without filaments in 40% of cases.

Culture

Culture at 37°C yields yeast colonies in 90% of cases. Susceptibility testing is appropriate in cases of recurrent or recalcitrant infection.

Management

Treat with topical nystatin tablets for 14 nights. Alternatively, treat with a topical azole such as clotrimazole, miconazole, econazole, ketoconazole creams and pessaries for 6 nights.

Oral treatments include:

- fluconazole, single dose 150 mg
- itraconazole, two doses 200 mg 8 hours apart.

Cutaneous candidosis

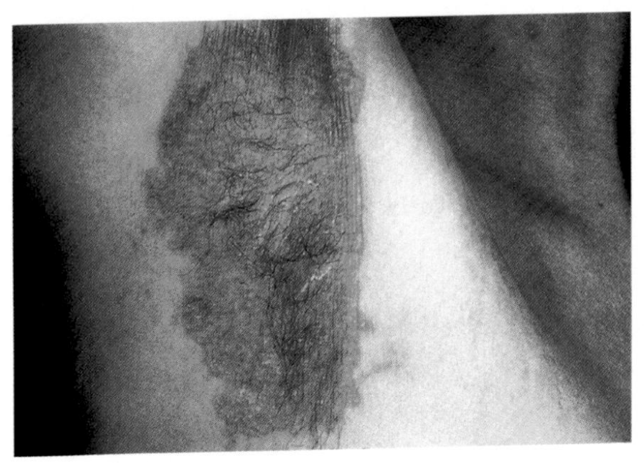

Candida albicans infection of axilla.

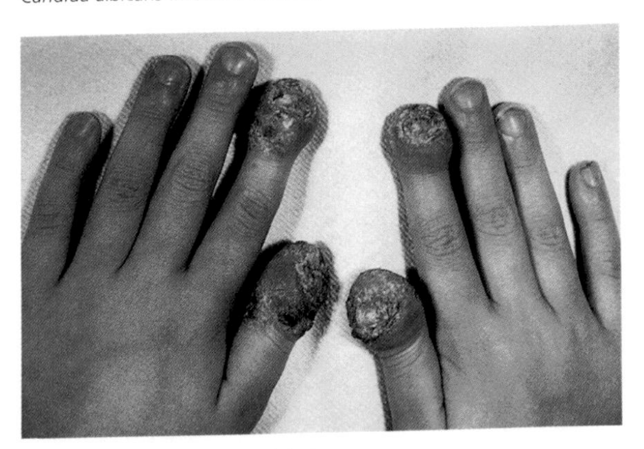

Chronic mucocutaneous candidosis.

Definition

Cutaneous candidosis is a yeast infection of the skin caused by members of the genus *Candida*. Infection of the proximal nail fold known as *Candida* paronychia may lead to nail infection.

Geographical distribution

World-wide.

Causal organisms and habitat

- Most commonly caused by *Candida albicans* then *C. tropicalis*, other species are occasionally implicated.
- Normal flora of the skin, mouth, intestinal tract and vagina.
- Disruption of the epidermal barrier function can lead to overgrowth and infection.

Clinical manifestations

- Affects intertriginous areas e.g. the groin, axillae and submammary folds.
- Erythematous, inflamed and painful skin.
- Predisposing factors include: coticosteroid, hormonal or antibiotic therapy, diabetes mellitus, obesity, maceration, friction, occlusion, immunosuppression, dermatitis, pregnancy, infancy or old age.
- *Candida* folliculitis may occur in occluded areas.
- *Candida* paronychia, infection of the proximal nail fold or cuticle, characterized by erythema, oedema and purulent discharge, may lead to nail infection.
- Chronic mucocutaneous candidosis is a severe, often widespread, erthematous or granulomatous infection of the mucous membranes, skin and nails seen in patients with congenital defects in their cell mediated immunity.

Essential investigations

Microscopy and culture

Microscopy of skin scrapings or swabs revealing yeasts with or without filaments is crucial in confirming infective status. Culture at 37°C yields yeast colonies after 24–48 hours.

Management

Oral agents are indicated for folliculitis, nail involvement, extensive lesions and in the immunocompromised.

- Itraconazole 200 mg/day or fluconazole 100 mg/day.

Control or eradication of the predisposing factor.

Topical therapy with azole agents, nystatin and naftifine, should be used twice daily until 1–2 weeks after clearing.

Additional steroid or antibacterial therapy may be indicated.

Malassezia infections

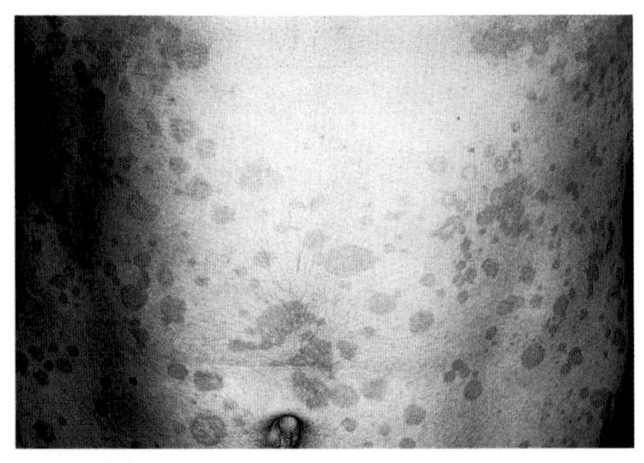

Pityriasis versicolor.

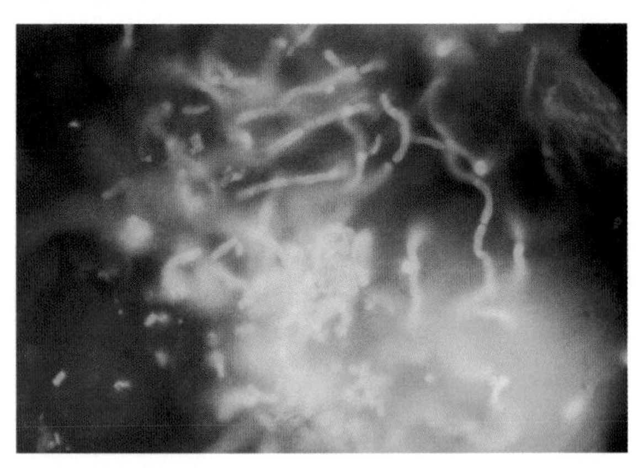

Microscopic appearance.

PITYRIASIS VERSICOLOR

Definition

Common, mild, often recurrent infection of the stratum corneum due to *Malassezia* yeasts.

Geographical distribution

World-wide, but it is more prevalent in tropical and subtropical regions.

In temperate climates, it is most common during the summer months.

Causal organism and habitat

- *Malassezia furfur* (previously known as *Pityrosporum orbiculare* and *P. ovale*).
- Budding yeast cells and numerous filaments.
- Normal skin of head and trunk.
- Sebaceous areas.
- Most common in adults 20–40 years of age.
- Human–human transmission possible.

Clinical manifestations

- Patches of fine, brown scaling on upper trunk, neck, upper arms and abdomen.
- Light-skinned patients: lesions initially pink then pale brown.
- Dark-skinned patients: skin loses colour, becomes depigmented.
- Most cases produce pale yellow fluorescence under Wood's light.

Essential investigations

Microscopy

Microscopy reveals oval, budding yeasts and short filaments.

Culture

A lipid supplement is required. When cultured at 32–34°C, small, yellow-cream colonies develop within 1 week.

Management

Good response to topical treatment with selenium sulphide (2%) shampoo, or ketoconazole shampoo or terbinafine.

Variable response to oral treatment with:

- ketoconazole 200 mg/day for 1 week
- itraconazole 200 mg/day for 1 week.

MALASSEZIA (*PITYROSPORUM*) FOLLICULITIS

Clinical manifestations

There are three main forms:

- Folliculitis on back or upper chest of young adults – scattered, itching follicular papules or pustules. These often appear after sun exposure.
- Associated with seborrhoeic dermatitis – numerous small follicular papules appear over upper and lower chest and back. There may be a florid rash, particularly marked on the back.
- In AIDS – multiple pustules on trunk and face, associated with severe seborrhoeic dermatitis.

Management

For extensive or recalcitrant lesions treat with

- oral ketoconazole 200 mg/day for 1–2 weeks.

Otherwise,

- topical imidazoles; or
- selenium sulphide.

SEBORRHOEIC DERMATITIS

Causal organisms and habitat

- Strong association with large numbers of *Malassezia* yeasts.
- Affects 2–5% of population.
- More frequent in men than in women.

Clinical manifestations

- Dandruff.

- Erythematous rash with scaling on scalp, face, ears, chest, and upper part of back:
  - scaling of eyelid margins and around nasal folds
  - in AIDS, onset is early sign of CD4 suppression
  - mycological investigation not required.

Management

Treat with topical imidazoles or mild corticosteroid creams. Ketoconazole shampoo is very effective in seborrhoeic dermatitis and dandruff. It should be applied twice per week for 2–4 weeks, then used at 1- or 2-week intervals to prevent recurrence. The combination of an azole with hydrocortisone is effective.

Mould infections of nails

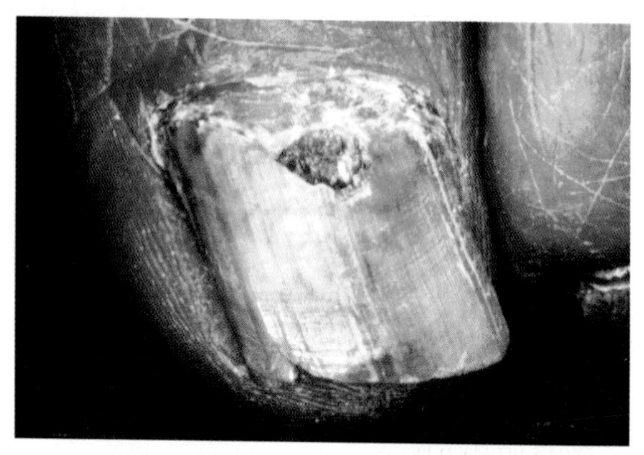

Nail showing onychomycosis due to *Scopulariopsis brevicaulis*.

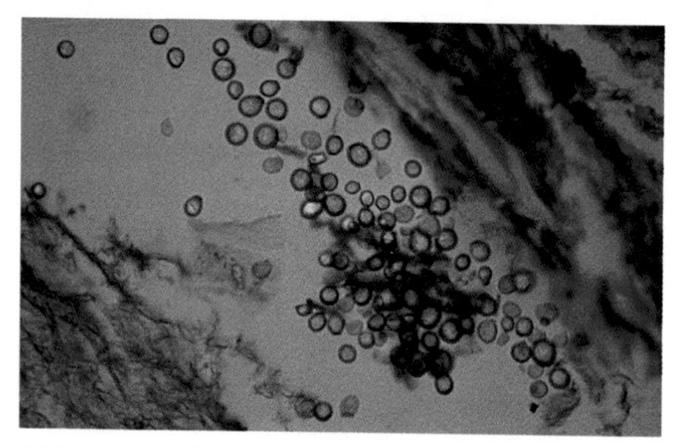

KOH preparation of a nail infected with *Scopulariopsis brevicaulis*.

Definition

The term onychomycosis is used to describe infection of the nails with fungi. In addition to the dermatophytes and *Candida* spp. there are a group of dermatomycotic moulds that can invade nail tissue.

Geographical distribution

World-wide.

Causal organisms and habitat

- Saprophytic moulds account for about 5% of nail infections.
- *Scopulariopsis brevicaulis* is implicated most frequently and can infect otherwise healthy nails.
- *Acremonium* spp., *Aspergillus* spp. (particularly *A. sydowii*/*A.versicolor*), *Fusarium* spp. and *Penicillium* spp. and more recently *Onychocola canadensis* are also encountered.
- *Scytalidium dimidiatum* (previously *Hendersonula toruloidea*) and *S. hyalinum* are encountered in patients of tropical origin.

Clinical manifestations

- Moulds usually only infect diseased or traumatized nails.
- Frequently only one nail is affected; toenails more commonly than fingernails; and it is seen in males more often than females, especially those over 50 years of age.
- No specific clinical features; the nail becomes lustreless and thickened. Small pits and streaks may appear in the nail plate, which is at first white, then yellow, brown, green or black.

Essential investigations

Microscopy and culture

The fungus must be seen on direct microscopic examination and grown in pure culture from most tissue samples. Cycloheximide-free medium should be used or mould growth will be suppressed.

Management

There is no consistent treatment of proven efficacy; however, chemical avulsion may be helpful.

- Occasionally individual cases respond to oral itraconazole 200 mg b.i.d. for 3–4 months or longer; or
- oral terbinafine 250 mg/day for 3–4 months; or
- topical amorolfine or tioconazole.

Treatment may have to be continued for 6 months or more.

Keratomycosis

Corneal ulcer due to *Aspergillus glaucus*.

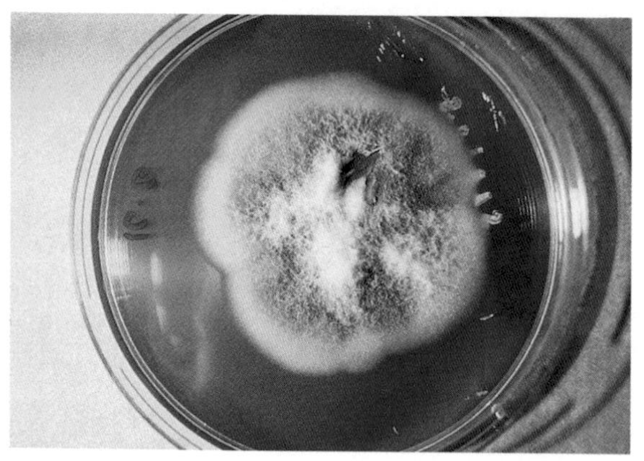

Culture of *Fusarium oxysporum*.

Definition

Keratomycosis is also referred to as mycotic keratitis and describes fungal infection of the cornea. Infection usually follows traumatic implantation of spores.

Geographical distribution

World-wide, but it is more common in the tropics.

Causal organisms and habitat

- Mainly due to saprophytic moulds (more than 60 species implicated).
- Most common are *Fusarium* spp., *Aspergillus* spp., *Curvularia* spp. and *Penicillium* spp. *Candida* spp. are also implicated.
- Infection follows traumatic injury either by direct implantation or subsequent infection of superficial abrasion.
- Topical antibiotics or steroids are predisposing factors.

Clinical manifestations

- Manifestations similar regardless of organism, although *Fusarium* spp. can produce toxins which increase local damage.
- Insidious in onset: increasing pain, ocular redness, photophobia and blurred vision.
- Slit lamp examination reveals: corneal ulcer with ragged white border, deep infiltrates, often discrete satellite lesions.

Essential investigations

Microscopy

Microscopy of corneal scrapings reveals fungal elements.

Culture

Culture at 28°C for 2 weeks to isolate causal organism. Significance increases with repeated isolation.

Management

Removal of infected tissue, discontinuation of steroids and topical or oral antifungal agent.

Treat with topical solutions such as 5% natamycin, 0.15% amphotericin B or 1% azole (econazole and miconazole). Topical treatment should be applied at hourly intervals for the first week. Thereafter it should be applied at similar intervals when the patient is awake.

Oral: fluconazole for yeasts, itraconazole for moulds. Six weeks of treatment is necessary for yeast and 12 weeks for mould infections.

Otomycosis

Culture of *Aspergillus niger*.

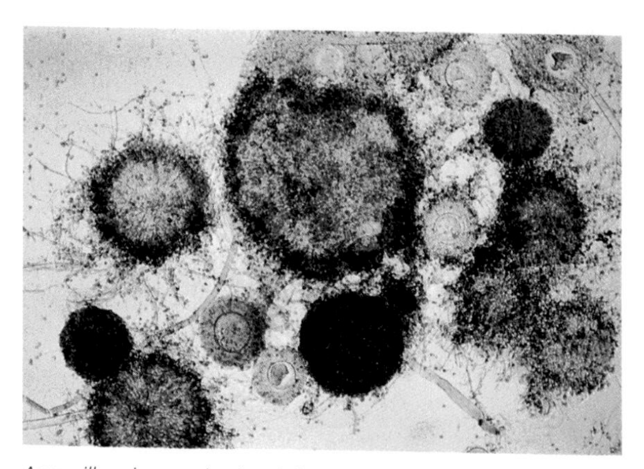

Aspergillus niger sporing heads from infected ear debris.

Definition

The term otomycosis describes infections of the ear canal.

Geographical distribution

World-wide but it is more common in warm climates.

Causal organisms and habitat

- Ubiquitous saprophytic moulds particularly *Aspergillus* spp. are implicated most frequently. *Scedosporium* spp., *Penicillium* spp., *Absidia* spp., *Rhizopus* spp., *Acremonium* spp. and *Scopulariopsis* spp. have also been reported.
- *Candida* spp. particularly *C. albicans* and *C. tropicalis* are also frequently isolated.
- In temperate regions it is most frequent during the summer.
- Often there is pre-existing aural disease.
- Topical antibiotics and steroids are predisposing factors.
- High humidity and waxy secretions favour mould growth.

Clinical manifestations

- Discomfort and irritation around the ear canal; there may be some hearing loss, tinnitus and giddiness.
- Discharge may be present; with mixed bacterial infections pain and suppuration is common.
- In advanced cases the mould can occupy much of the lumen of the canal and may be visible as a woolly mycelial mat.

Essential investigations

Microscopy and culture

Microscopic examination of debris from the ear canal will reveal branching hyphae, budding cells or both.

Sporing heads may be seen in *Aspergillus* spp. infections.

Isolation in culture at 28°C allows identification of the infecting organism.

Management

Treatment commences with thorough cleaning followed by the application of an antifungal.

- Topical natamycin or nystatin morning and evening for 2–3 weeks. Local application of an imidazole cream, such as clotrimazole or econazole nitrate, also gives good results.
- Insertion of a regularly replaced gauze pack soaked in amphotericin B, natamycin or an azole for 1 week.

Aspergillosis

Aspergillus flavus and *Aspergillus fumigatus* in culture.

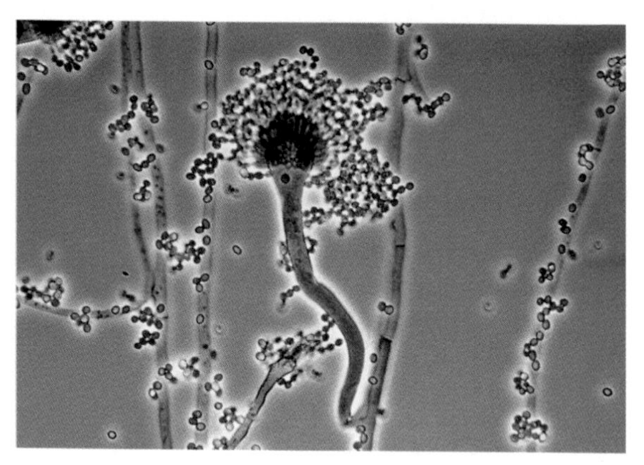

Aspergillus fumigatus sporing head.

Definition

The term aspergillosis describes infections due to moulds belonging to the genus *Aspergillus.* These can range from localized infections to life-threatening systemic infection. Disease may also result from an allergic reaction to inhaled spores.

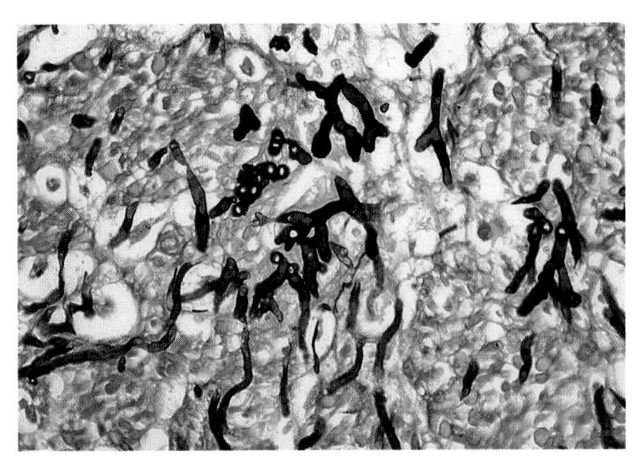

Histopathological appearance of *Aspergillus* lung infection.

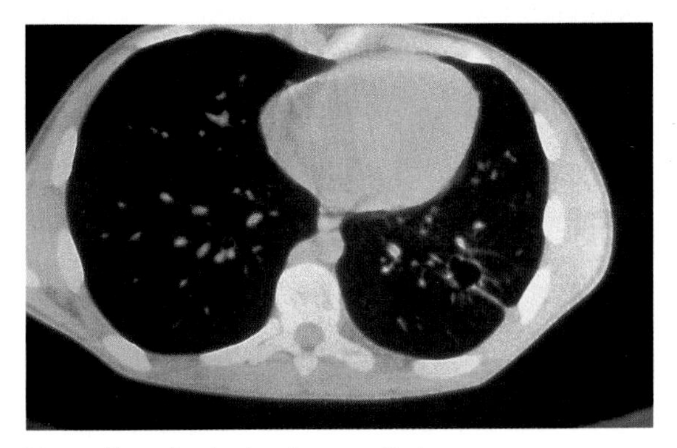

CT scan of lung showing invasive aspergillosis.

Geographical distribution

World-wide.

Causal organisms and habitat

Aspergillus species found:

- in soil, air, plants and decomposing organic matter
- in dust and on food in homes
- in hospital environments, especially ceiling voids, foods, plants, fabrics and in association with demolition or construction work, or defective ventilation.

Clinical manifestations

Allergic aspergillosis

- Uncommon, but most often seen in atopic individuals.
- Bronchial allergic reaction.
- Mucus plugs form in bronchi, leading to atelectasis.
- Often progresses to bronchiectasis and fibrosis.
- Results from type I and III reaction to *Aspergillus* antigens.
- Fever, intractable asthma, productive cough, malaise, weight loss.
- Expectoration of brown eosinophilic mucus plugs containing *Aspergillus* is common.
- Variable radiographic findings.

Aspergilloma (fungus ball)

- Occurs in patients with residual lung cavities following tuberculosis, sarcoidosis, bronchiectasis, pneumoconiosis and ankylosing spondylitis.
- Patients often asymptomatic.
- Chronic cough.
- Malaise, resulting in weight loss.
- Haemoptysis in 50–80% of cases, in 25% can be life threatening.
- Upper lung lobes:
  - oval or round mass with radiolucent halo or air crescent on radiographs
  - CT will delineate lesion.
- Spontaneous lysis in 10% of cases.

Chronic necrotizing aspergillosis of the lung

- Indolent condition seen in middle-aged or older patients with underlying lung disease.
- Seen associated with alcoholism or diabetes in particular.
- Often similar in clinical presentation to aspergilloma.
- Most frequent symptoms are fever, a productive cough, malaise and weight loss.
- Radiological changes:
  - chronic upper lobe infiltrate associated with pleural thickening
  - cavitation is common.

Acute invasive pulmonary aspergillosis

- Occurs in compromised patients and is often fatal.
- Patients at risk include neutropenic cancer patients, transplant recipients, patients with GVHD, AIDS patients, children with CGD. It may be a focal or diffuse infection, with haematogenous dissemination a frequent complication.
- In neutropenic patients, symptoms include unremitting fever that fails to respond to broad-spectrum antibiotics, pleuritic chest pain and coughing. Typical CT chest signs are small nodular lesions, larger peripheral lesions and a halo sign around nodular lesions. However, signs of diffuse infection are much less distinctive.

Tracheobronchitis and obstructing bronchial aspergillosis

- Tracheobronchitis:
  - AIDS and lung transplant recipients
  - dyspnoea and wheezing
  - CT scan ineffective
  - diagnosis established by bronchoscopy.
- Obstructing bronchial aspergillosis:
  - noninvasive
  - seen most often in AIDS
  - cough, fever, wheezing
  - expectoration of large mucus plugs
  - if untreated, can become invasive
  - diagnosis established by bronchoscopy.

Sinusitis

- Most common form of fungal sinusitis.
- Five patterns of infection:
  - allergic sinusitis
  - acute invasive sinusitis
  - chronic necrotizing sinusitis
  - aspergilloma (fungal ball) of the paranasal sinuses
  - paranasal granuloma.

Cerebral aspergillosis

- Follows haematogenous dissemination from lungs.
- 10–20% brain involvement.
- Seldom diagnosed during life.

- *Aspergillus* common cause of brain abscesses in bone marrow transplant recipients.
- Disease progression gradual in onset.
- CT helpful in locating lesions, but findings are nonspecific.

Ocular aspergillosis

Three forms:
- corneal infection
- endophthalmitis
- orbital infection.

Endocarditis

- Mainly associated with heart transplantation.
- Complication of parenteral drug abuse.
- Similar symptoms to bacterial endocarditis:
  - fever, weight loss, fatigue, loss of appetite
  - murmurs in 50–90% of patients.

Cutaneous aspergillosis

Two forms in immunocompromised patients:
- primary infection at catheter insertion sites due to contaminated splints
- haematogenous spread to skin in about 5% of patients with invasive aspergillosis.

Aspergillosis in AIDS

- Seen in about 4% of patients.
- Lung most common site.
- Bronchoscopic examination helpful.

Essential investigations

Microscopy

Sputum examination is helpful in ABPA, but of limited use in invasive disease.

However, BAL is sometimes helpful in invasive disease. Histopathology is the most reliable diagnostic method, though a similar appearance is seen with *Fusarium* and *Scedosporium*.

Culture

Culture provides the definitive diagnostic method, although interpretation is difficult.

Isolation may come from sputum and BAL.

Aspergillus is seldom isolated from blood, urine or CSF, but can often be isolated from sinus washings or biopsies.

Clinical tests

CT and MRI scans have enhanced the early diagnosis of invasive aspergillosis as lesions are often visible earlier than on a chest X-ray. Diagnostic signs include a halo presentation and the appearance of crescent-shaped air pockets.

Serology

Precipitin testing is useful in ABPA and aspergilloma.

Antigen testing is useful in invasive disease, such as ELISA for galactomannan (Sanofi Platelia *Aspergillus*).

PCR may be used for the detection of fungal genomic sequences (experimental).

Management

Allergic aspergillosis

Mild disease does not require treatment, but when treatment is indicated, prednisolone is drug of choice:

- 1.0 mg/kg/day
- when radiographs clear 0.5 mg/kg/day for 2 weeks
- 0.5 mg/kg at 48-hour intervals for 3–6 months.

Aspergilloma

- Surgical removal of the lesion is necessary.
- Endobronchial instillation or percutaneous instillation of amphotericin B 10–20 mg in 10–20 ml distilled water instilled two or three times per week for about 6 weeks.

Chronic necrotizing aspergillosis

- Itraconazole 200–400 mg/day.
- Surgical resection of necrotic lung with local amphotericin B.

Sinusitis

- Allergic:

   - prednisolone.
- Noninvasive:
   - surgical resection.
- Invasive:
   - surgical debridement
   - amphotericin B 1 mg/kg/day
   - AmBisome 3–5 mg/kg/day
   - itraconazole 400–600 mg/day.

Paranasal granuloma

Surgical debridement and itraconazole 200–400 mg/day.

Acute invasive pulmonary aspergillosis

- Prompt treatment.
- Variable response rate:
   - 10% in BMT recipients
   - 30% in neutropenic patients.
- Amphotericin B 1.0–1.5 mg/kg/day.
- AmBisome 3–5 mg/kg/day or higher.
- Amphocil (Amphotec) 3–4 mg/kg/day or up to 6 mg/kg/day.
- Abelcet 5 mg/kg/day.
- Itraconazole 400–600 mg/day for 4 days then 200 mg twice daily.

Cerebral

- Poor prognosis.
- AmBisome 3–5 mg/kg and higher.

Cutaneous

- Amphotericin B 1.0 mg/kg/day.
- Surgical debridement at catheter insertion sites.

Prophylaxis

- Itraconazole oral solution 400 mg/day.
- Amphotericin B 0.5 mg/kg/day.

Empirical

- Amphotericin B 1 mg/kg/day.
- AmBisome 3 mg/kg/day.

Pneumocystis carinii pneumonia

Definition

Infection with *Pneumocystis carinii* usually presents as a pnuemonitis (PCP). It occurs in immunosuppressed or debilitated patients and is the commonest cause of pneumonia in AIDS.

Causal organism and habitat

World-wide distribution. *Pneumocystis carinii* shares morphological and structural features with both fungi and protozoa.

Clinical manifestations

- The lung is the primary site of infection.
- Tachypnoea may be the only sign.
- Cysts are formed in lungs and other tissues, especially kidneys; spread is probably haematogenous.
- AIDS is the defining illness in 30% of cases.

Essential investigations

- *P. carinii* cannot be isolated in culture; diagnosis is by detection of cysts or 'trophozoites'.
- Immunofluorescence staining with specific monoclonal antibodies.
- The yeast forms of *P. carinii* may resemble *Histoplasma capsulatum* but they do not bud and are usually extracellular.

Management

P. carinii lacks ergosterol in its cell membrane and responds better to antibacterial and antiprotozoal than antifungal agents.

Co-trimoxazole 120 mg/kg/day in divided doses for 2–3 weeks.

An alternative is pentamidine isethionate 4 mg/kg/day for 2–3 weeks.

Very high mortality despite treatment.

Prophlyaxis is indicated in some patient groups.

Deep candidosis

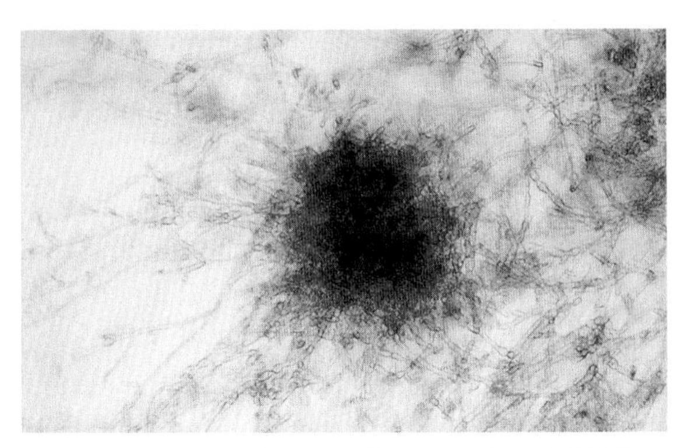

Microscopic appearance of *Candida albicans* in cerebrospinal fluid.

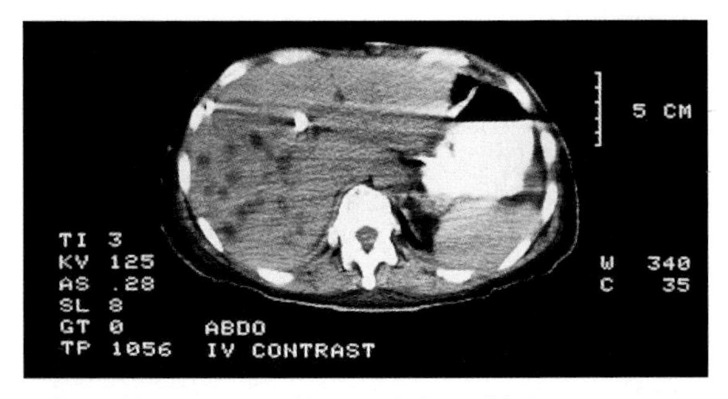

Radiographic appearance of hepatosplenic candidosis.

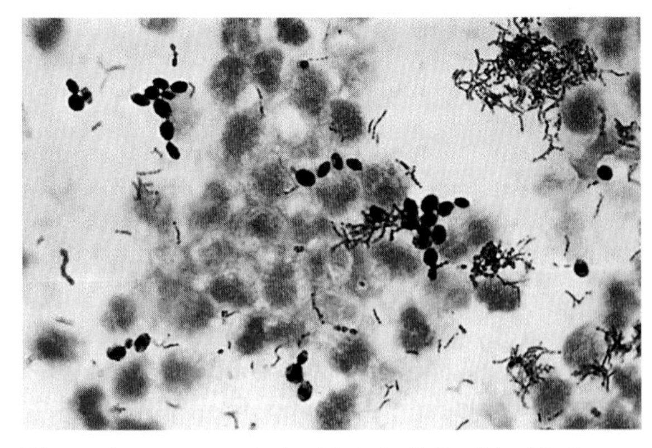

Microscopic appearance of urine infected with *Candida glabrata*.

Gross pathology of liver infected with *Candida albicans*.

Definition

Acute or chronic deep-seated infection due to organisms belonging to the genus *Candida*. Infections are seen in surgical, debilitated or immunocompromised patients and may be confined to one organ or become systemic.

Geographical distribution

World-wide.

Causal organisms and habitat

- *Candida albicans* remains most frequent cause.
- Other species, particularly *C. tropicalis, C. parapsilosis* and *C. glabrata* increasing in prevalence. Many others implicated less frequently.
- *Candida* spp. part of normal flora of skin, mouth, intestinal tract and vagina thus breaches of epidermal barrier function can lead to infection.
- Most deep-seated infections endogenous in origin.
- Outbreaks due to transmission on hands of healthcare workers reported in neonatal, surgical intensive care units and burns units.
- Risk factors include: surgery; neutropenia or other immunosuppression; intensive care; trauma; malnutrition; extensive antibiotic therapy; adrenal corticosteroids; catheterization; prematurity and low birth weight; and intravenous drug abuse.

• Candidaemia increasing in incidence; recorded as the fourth most common bloodstream infection.

Clinical manifestations

Acute disseminated candidosis

• Disseminated infection occurs more frequently than single organ infection.
• Nonspecific clinical signs and symptoms; infection should be suspected in any patient with one or more predisposing factors who develops a fever unresponsive to broad-spectrum antibiotics.
• Endophthalmitis occurs in 5–50% of patients with disseminated candidosis but rarely seen in neutropenic patients who develop the disease.
• Characteristic macronodular skin lesions occur in up to 10% of patients.
• Can involve any organ, typically infection extends to three or four organs, most commonly the kidneys (80%), heart, gastrointestinal tract and lung.
• Pattern of invasion suggests initial entry into the systemic circulation.

Chronic disseminated candidosis

• Previously known as hepatosplenic candidosis, frequently involves the liver and spleen, often seen in leukaemic patients who have recovered their neutrophil counts.
• Pattern of infection suggests initial invasion via the portal circulation from the GI tract.

Congenital candidosis

• Rare condition which can result in fetal or neonatal death.
• Ascending maternal infections linked to symptomatic vaginal candidosis (the incidence of which increases to 25% during pregnancy), the presence of intrauterine devices, cervical sutures, antibiotic use and premature rupture of the membranes.
• Occasionally yeasts may be transferred to the fetus via invasive procedures such as amniocentesis.

Oesophagitis

- Symptoms include oesophadynia and dysphagia.
- Occurs in AIDS patients or following cancer chemotherapy.
- Most commonly seen in patients with oral candidosis.

Gastrointestinal candidosis

- Often asymptomatic.
- Pre-existing mucosal ulceration may be superinfected:
  - perforation can lead to dissemination
  - intestinal candidosis in patient groups other than those undergoing cancer chemotherapy or those with AIDS remains controversial.

Urinary tract and renal candidosis

- Symptoms include fever, rigors, and lumbar and abdominal pain.
- Ascending renal *Candida* infections may be due to catheterization or instrumentation of the bladder or local spread of infection from the perianal area.
- Yeasts may be filtered from the blood thereby infecting the kidney, 80% of patients with disseminated candidosis develop renal candidosis.
- Fungal ball may form in the kidney obstructing the renal pelvis leading to anuria.

Pulmonary candidosis

- Uncommon infection in debilitated or neutropenic patients.
- Haematogenous spread or endobronchial inoculation.
- Nonspecific clinical and radiological presentation.

CNS candidosis

- *Candida* meningitis is uncommon.
- Low birth-weight infants and neurosurgical patients.
- Haematogenous spread or direct inoculation.
- CSF findings indistinguishable from bacterial meningitis and include raised protein and decreased glucose concentrations.
- Brain abscess may be diagnosed by CT or MRI scan.
- Diffuse metastatic encephalitis is rarely diagnosed during life.

Endocarditis, myocarditis and pericarditis

- Symptoms indistinguishable from those of bacterial endocarditis and include: fever, weight loss, fatigue, heart murmurs and enlarged spleen.
- Risk factors include: damaged or prosthetic heart valves and intravenous drug abuse.
- Indolent onset often with a silent period of several weeks or months.
- Vegetations may be apparent on echocardiogram.
- Myocardial infection with abscess formation may be a consequence of endocarditis or arise as a result of haematogenous dissemination.
- Pericarditis with associated chest pain, pericardial friction rub and effusion is most frequently a complication of a superficial myocardial abscess.

Peritonitis

- *Candida* peritonitis associated with fever, abdominal pain and tenderness may be an infective complication of peritoneal dialysis or may occur as a result of gastrointestinal perforation or leaking intestinal anastomosis.

Endophthalmitis

- Painful condition resulting from haematogenous dissemination.
- Fluffy yellow-white exudates may be seen on the retina. These are due to the fungus and the consequent inflammatory reaction which may explain their rarity in neutropenic patients.
- May be a consequence of intravenous heroin abuse.

Osteomyelitis, arthritis and myositis

- Characterized by localized pain but rarely fever.
- Usually arise as a consequence of haematogenous dissemination, less commonly traumatic implantation.

Catheter associated candidaemia

- Intravenous catheters become infected with yeasts leading to positive blood cultures or candidaemia.
- Infection may clear on removal of the catheter or have progressed to disseminated candidosis.

Essential investigations

Microscopy and culture

Microscopy and culture of normally sterile body fluids reveals yeast cells with or without filament production, and growth of yeast colonies after 24–48 hours.

Isolates should be speciated and where appropriate antifungal susceptibility testing carried out.

Histopathology provides definitive evidence.

Blood culture may result in a diagnosis of candidaemia. This may be line-associated and transient, or indicative of disseminated disease.

Clinical tests

CT and MRI scans may be helpful in diagnosing chronic disseminated candidosis (a normal liver scan has a high negative predictive value) and determining the extent of brain involvement.

Fundoscopic examination may reveal opacities, whilst electrocardiograms may be useful in detecting heart valve vegetations.

Serology

Serology is useful in immunocompetent individuals; high or rising antibody titres (1 : 8 or greater) are considered indicative of active infection, a raised antibody titre is the single most consistent finding in *Candida* endocarditis. Quantitative determination of anti-*Candida* mannan IgG is useful in various manifestations (Sanofi Platelia *Candida* antibody).

Detection of mannoprotein antigen (Sanofi Platelia *Candida* antigen) by ELISA is useful.

Experimental PCR has been used to detect yeast genomic sequences.

Management

Selection of therapy should be guided by the species of yeast isolated and antifungal susceptibility test results where appropriate.

Candida oesophagitis

Effective treatments for *Candida* oesophagitis include these alternatives:

- Oral ketoconazole 200–400 mg/day for 1–2 weeks.
- Fluconazole 100–200 mg/day for 1–2 weeks.
- Itraconazole 200–400 mg/day for 2 weeks.

Deep and disseminated infection

Documented candidaemia should always be treated and lines should be removed or replaced where possible.

Disseminated infection and most deep forms of candidosis should be treated with :

- Amphotericin B 1 mg/kg/day for 4–6 weeks with or without the addition of flucytosine 100–200 mg/kg/day in four divided doses depending on the organism's susceptibility to this agent.
- Depending on the infecting species, fluconazole 200–400 mg/day may also be effective.

Fluconazole is excreted unchanged in the urine so is useful for urinary tract infections and also reaches high concentrations in the vitreous humour so may be useful in *Candida* endophthalmitis.

Lipid preparations of amphotericin B should be considered in patients who fail to respond to, or develop side-effects to the conventional formulation.

In cases of chronic disseminated candidosis

- Liposomal amphotericin (AmBisome) 3–5 mg/kg/day is especially useful because it accumulates in the liver.

Candida endocarditis

Removal of infected valves is the treatment of choice for *Candida* endocarditis, antifungal cover with a combination of amphotericin B and flucytosine should be given prior to surgery and should be continued for 2–3 months to help prevent relapse.

Prophylaxis

Prophylactic chemotherapy with either

- fluconazole 100 mg/day or
- itraconazole 100–200 mg/day should be considered in patients at risk from invasive yeast infection.

Empirical

Empirical therapy with amphotericin B or one of its lipid formulations should be considered in neutropenic patients with persistent fever despite 96 hour therapy with broad-spectrum antibiotics.

- Amphotericin B 1.0 mg/kg.
- AmBisome 1–3 mg/kg/day.

Cryptococcosis

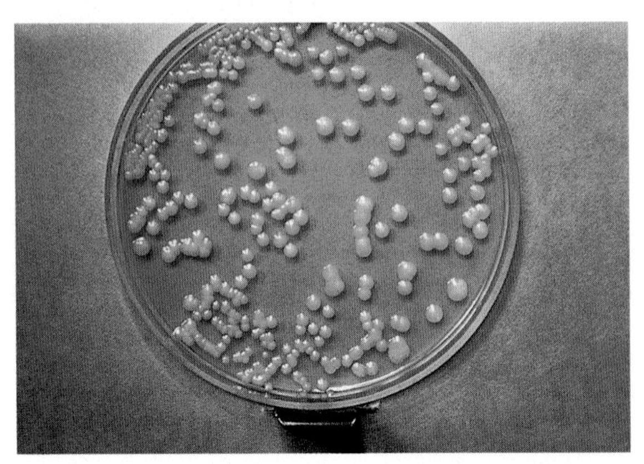

Appearance of *Cryptococcus neoformans* on Niger seed agar.

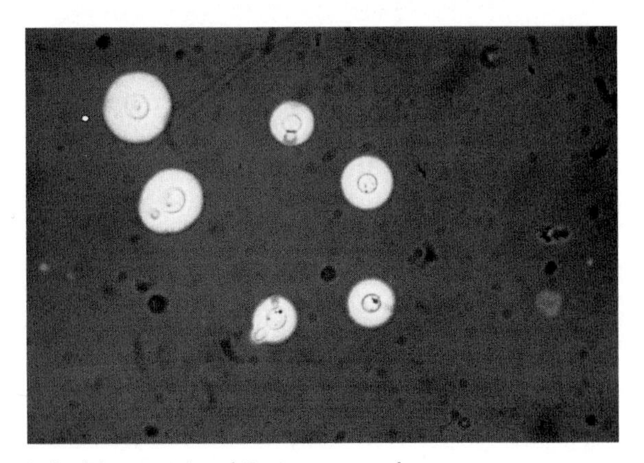

Indian ink preparation of *Cryptococcus neoformans*.

Definition

Infection with the encapsulated yeast *Cryptococcus neoformans*.

Most infections occur in immunocompromised patients, especially those with AIDS.

Meningitis is the most common clinical presentation.

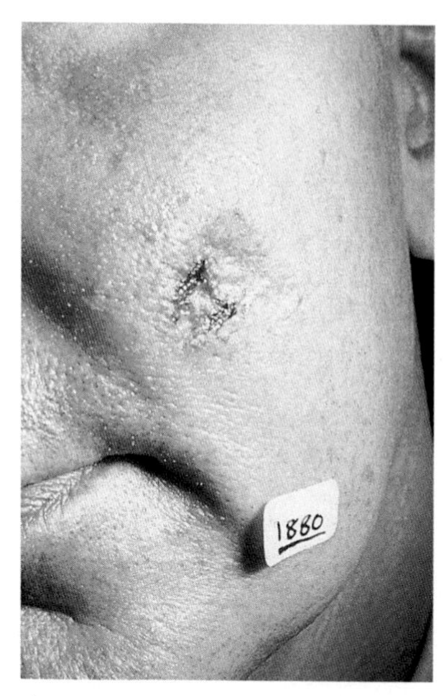

Cutaneous cryptococcosis.

Geographical distribution

World-wide, but there are differences in distribution of the causal species.

Causal organisms and habitat

- Encapsulated yeast.
- Two varieties:
  - *C. neoformans* var. *neoformans* serotypes A and D which are encountered world-wide, often in pigeon guano.
  - *C. neoformans* var. *gattii* serotypes B and C are restricted to tropical and subtropical regions, often in association with the red gum tree.
- Sexual form is the filamentous mould *Filobasidiella neoformans*.

Clinical manifestations

- Infection is believed to follow inhalation of desiccated spores.

- Pulmonary infection is asymptomatic in 30% of normal individuals, others have a productive cough, chest pain, weight loss and fever.
- Radiology reveals well-defined, non-calcified, single or multiple nodular lesions.
- Pulmonary infection may occur and resolve weeks to months before disseminated infection becomes manifest in compromised patients.
- Immunocompromised patients often present with meningitis.
- T-cell mediated immunological defects are the main predisposing factor.
- Major cause of morbidity and death in AIDS patients.
- Abnormal CSF findings:
  - raised pressure, increased protein, lowered glucose
  - lymphocytic pleocytosis.
- Often insidious in onset
- Cutaneous, osteomyelitic or endophthalmitic lesions may follow haematogenous spread of the organism.
- The prostate is a reservoir for relapse.
- Infection is usually with *C. neoformans* var. *neoformans* even in areas where *C. neoformans* var. *gattii* is found in the environment.

Essential investigations

Microscopy

Microscopic examination of CSF mounted in Indian ink reveals encapsulated yeast cells.

Culture

Culture of CSF, sputum, blood, urine, prostatic fluid; the sample should be centrifuged and the residue plated onto glucose peptone agar and incubated at 30–35°C for 2 weeks.

Positive blood cultures are found in 35–70% of AIDS patients.

Niger seed agar can be used to distinguish *Cryptococcus* spp. (brown colonies) from *Candida* spp. (white colonies).

Concanavalin medium can be used to distinguish *C. neoformans* var. *neoformans* (yellow) from *C. neoformans* var. *gatii* (blue).

Antigen test (LPA or ELISA) on CSF, serum, urine and BAL is a very reliable diagnostic test.

Ninety per cent of patients with cryptococcal meningitis have a positive LPA test, but titres are higher in AIDS patients.

Antibodies are detected in patients with early or localized infection.

Management

- A combination of amphotericin B 0.3–0.5 mg/kg/day increasing to 0.6–1.0 mg/kg/day if necessary with flucytosine 100–150 mg/kg/day for 4–6 weeks.
- Alternatively, amphotericin B can be administered alone (0.8–1.0 mg/kg/day) for 10 weeks.
- AIDS patients may respond to a 2-week course of combination therapy followed by 400 mg/day fluconazole.

AIDS patients require life-long maintenance with fluconazole 200 mg/day.

Antigen titres should be monitored at 1, 2, 3 and 6 months following treatment but they should be interpreted with caution. A change of less than four-fold is not considered significant and a high titre is not necessarily indicative of active infection. Titres on CSF and serum should only be considered in conjunction with a clinical evaluation. Serum titres are usually lower than CSF titres.

Successful treatment includes resolution of clinical symptoms and two consecutive negative cultures.

It can be considered relapse rather than recalcitrant infection if the patient is free from clinical signs and symptoms and previously positive cultures are negative for a period of several months.

Reinfection from an environmental source is also possible.

Mucormycosis

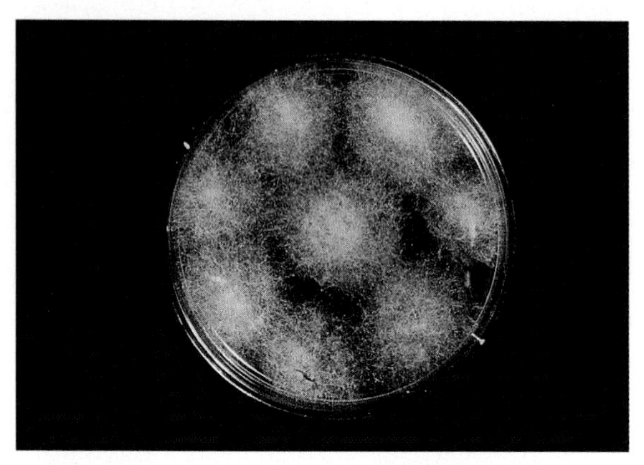

Culture of *Absidia corymbifera*.

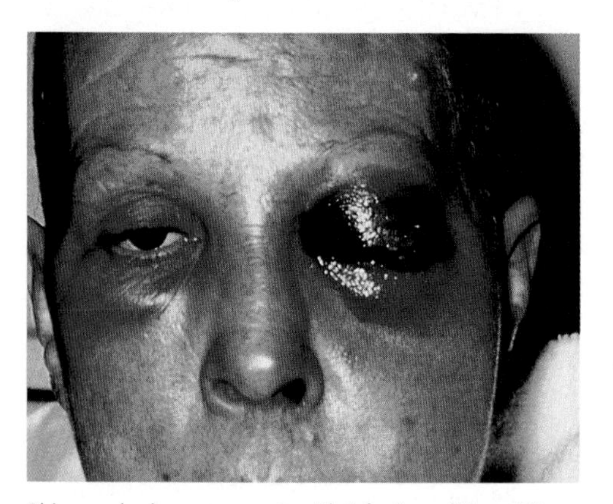

Rhinocerebral mucormycosis with infection of the orbit.

Definition

Mainly rhinocerebral, but also pulmonary, gastrointestinal cutaneous or disseminated infection caused by moulds belonging to the order Mucorales.

Geographical distribution

World-wide.

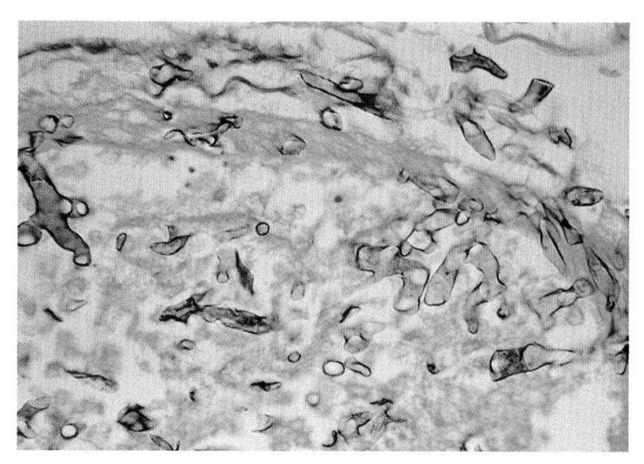

Histopathological appearance of mucormycosis.

Causal organisms and habitat

- *Rhizopus arrhizus.*
- *Absidia corymbifera.*
- *Apophysomyces elegans.*
- *Cunninghamella bertholletiae.*
- *Rhizomucor pusillus.*
- *Saksenaea vasiformis.*
- Ubiquitous in enviroment – in soil, food and on decomposing organic matter:
  - spores inhaled
  - less frequently, traumatic inoculation into skin.

Clinical manifestations

- Opportunistic infection.
- Clinical forms associated with particular underlying disorders.
- Predilection for vascular invasion:
  - thrombosis
  - infarction
  - tissue necrosis.
- Classic feature is rapid onset of necrosis and fever.
- Rapid progress and death if not treated aggressively.

Rhinocerebral mucormycosis

- Starts in paranasal sinuses.
- Spreads to orbit, face, palate and/or brain.
- Most commonly seen associated with uncontrolled diabetes.
- Often fatal if left untreated.
- Initial symptoms include unilateral headache, nasal or sinus congestion, serosanguinous nasal discharge and fever:
  - spreads into palate and forms black necrotic lesion
  - nasal septum or palatal perforation frequent.
- Consequences of spread into orbit:
  - periorbital or perinasal swelling occurs
  - induration and discoloration
  - ptosis and/or proptosis
  - loss of vision
  - drainage of black pus
  - angioinvasive spread to brain is common.
- CT and MRI useful in defining extent of bone and soft-tissue destruction.

Pulmonary mucormycosis

- Most cases in neutropenic cancer patients.
- Nonspecific symptoms include fever and cough.
- Seldom diagnosed during life.
- Aspiration of infectious material.
- Inhalation.
- Haematogenous spread during dissemination.
- Fatal within 2–3 weeks.

Gastrointestinal mucormycosis

- Uncommon.
- Malnourished infants or children.
- Stomach, colon and ileum.
- Seldom diagnosed during life.
- Varied symptoms, typically nonspecific abdominal pain and haematemesis.
- Peritonitis if intestinal perforation occurs.

Cutaneous mucormycosis

- Associated with burns: spread to underlying tissue.
- Severe underlying necrosis develops.
- In diabetics: cutaneous lesions at injection site.

- Associated with contaminated surgical dressings or splints.

Disseminated mucormycosis

- Develops from other manifestations.
- Usually in neutropenic patients with pulmonary infection.
- Brain is most common site of spread.
- Metastatic lesions found in spleen, heart and other organs.
- Seldom diagnosed during life.
- Occasional cutaneous lesions permit early diagnosis.

Cerebral infection alone

- Follows haematogenous dissemination.
- Distinct from rhinocerebral mucormycosis.
- Focal neurological signs.
- Difficult diagnosis:
  - neutropenic patient
  - intravenous drug abusers
  - confusion, obtunded or somnolent
  - CT and MRI useful but nonspecific
  - CSF investigations unhelpful.

Essential investigations

Microscopy and culture

The presence of broad, nonseptate hyphae with right-angled branching in specimens from necrotic lesions, sputum or BAL is highly significant.

Nasal, palatal and sputum cultures are seldom helpful.

Interpretation should be cautious, but isolation should not be ignored if the patient is diabetic or immunosuppressed.

Management

Control the underlying disorders, including prompt correction of acidosis. Also, remove infected necrotic tissue. Treat with:

- amphotericin B 1.0–1.5 mg/kg/day, continued for 8–10 weeks, until reaching a total dose of 2 g
  - liposomal amphotericin B 3–5 mg/kg or higher
- surgical resection in pulmonary disease
- aggressive surgical debridement of necrotic lesions in cutaneous mucormycosis.

Blastomycosis

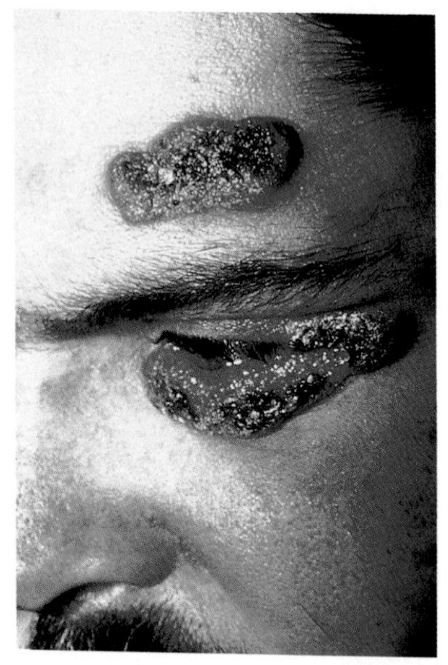

Cutaneous blastomycosis.

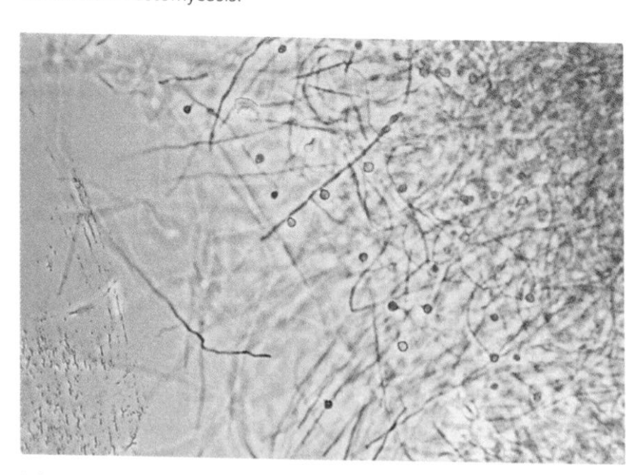

Microscopic appearance of *Blastomyces dermatitidis*.

Definition

Pulmonary infection caused by *Blastomyces dermatitidis* in

normal individuals but it often spreads to involve other organs, particularly skin and bones.

Geographical distribution

Midwestern and southeasten regions of North America; it also occurs in Central and South America and parts of Africa.

Causal organism and habitat

- *Blastomyces dermatitidis.*
- Exists in nature as mycelium; in tissue as large, round budding cells.
- Soil is natural habitat; greatest survival in moist soil containing organic debris or in rotting wood.
- Associated with outdoor occupations or recreational interest.
- Endemic regions not easily delineated.
- Much more common in men.
- Risk factors have not been identified.

Clinical manifestations

Pulmonary blastomycosis

- Infection follows inhalation.
- Lungs initial site of infection.
- Pulmonary lesions often not detected until infection has spread to other sites.
- Acute disease: flu-like illness: fever, chills, productive cough, myalgia, arthralgia, pleuritic chest pain.
- Radiological findings are nonspecific; include lobar or segmental consolidation, often in lower lobes.
- Most patients recover after 2–12 weeks of symptoms.
- Patients who fail to recover develop chronic pulmonary infection or disseminated disease.

Cutaneous blastomycosis

- Cutaneous lesions occur in over 70% of cases with disseminated disease.
- Painless raised verrucous lesions with irregular borders.

• Face, upper limbs, neck and scalp are most frequently involved.

Osteoarticular blastomycosis

• Occurs in about 30% of patients with disseminated disease.
• Spine, ribs, long bones most common sites of infection.
• Lesions often remain asymptomatic until infection spreads into contiguous joints, or into adjacent soft tissue causing abscess formation.
• Radiological findings nonspecific; well-defined osteolytic or osteoblastic lesions cannot be distinguished from other fungal or bacterial infections.
• Arthritis occurs in up to 10% of patients; in knee, ankle, elbow or wrist.

Genitourinary blastomycosis

• Prostate involved in 15–35% of men with disseminated blastomycosis.
• Epididymitis presents as scrotal swelling.

Blastomycosis in special hosts

• Occasionally associated with impaired T-cell mediated immunological function.
• Occasionally seen in AIDS.

Essential investigations

Microscopy

Large round cells with thick refractile walls and broad-based single buds seen in pus, sputum, bronchial washings and urine.

Culture

Culture provides the definitive diagnosis. Mycelial colonies are seen after 1–3 weeks at 25–30°C. Identification can be confirmed by subculture on brain–heart infusion agar at 37°C.

Serology

Complement fixation is insensitive and nonspecific.

Immunodiffusion is more specific but negative reactions occur in many patients with the disease.

Management

Many patients who are asymptomatic recover without treatment but may develop serious complications later; therefore monitor for signs of reactivation.

All patients with symptomatic infection require treatment.

- Oral itraconazole 200 mg/day for up to 6 months for non-immunocompromised patients with indolent forms of the disease. Continue for at least 3 months after the lesions have resolved.
- In nonresponders, or where there is progression, increase itraconazole to 400 mg/day initially then increase to 600–800 mg/day as required.

Fluconazole is useful where itraconazole is not absorbed.

In cases of life-threatening infection or where CNS is involved:

- Amphotericin B 0.3–0.6 mg/kg, to a total dose of 1.5–2.5 g. This may be used in immunocompromised patients or itraconazole failures. Switch to itraconazole after initial amphotericin B response.

Coccidioidomycosis

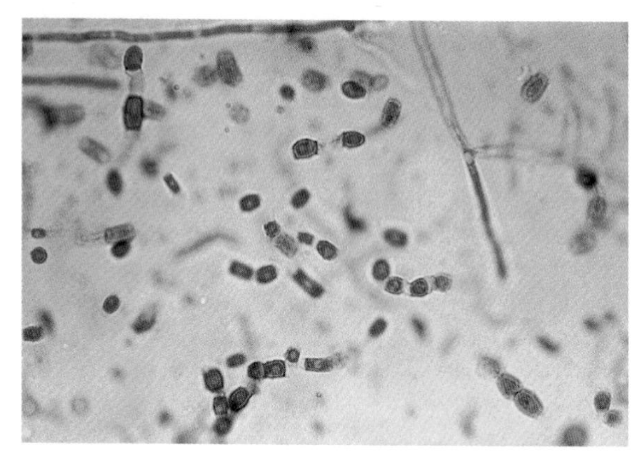

Microscopic appearance of *Coccidioides immitis* arthrospores.

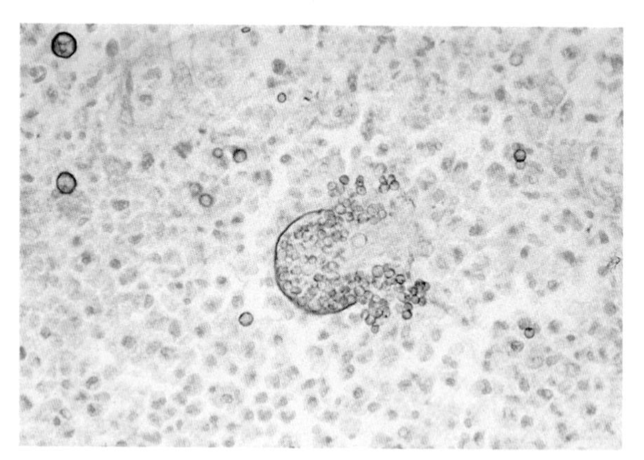

Histopathological appearance of *Coccidioides immitis* spherules in tissue.

Definition

A mild, transient pulmonary infection caused by the dimorphic fungus *Coccidioides immitis*. It can proceed to a progressive infection of the lungs or more generalized infection in immunosuppressed patients.

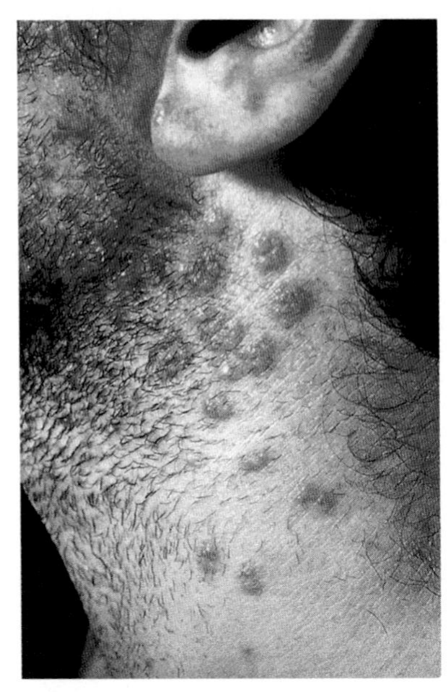

Erythema multiforma in a patient with primary pulmonary coccidioidomycosis.

Geographical distribution

Most cases occur in the southwestern USA, and in parts of Central and South America.

Infection diagnosed outside these regions occurs in individuals exposed in endemic regions.

Causal organism and habitat

- *Coccidioides immitis*, found in soil.
- Exists in nature as mycelium, which fragments into arthrospores.
- Forms large round thick-walled spherules containing endospores in tissue.
- Restricted geographical distribution: hot, arid regions of southwestern USA, parts of Central and South America.
- Dust storms often spread the organism far outside its endemic regions.

Clinical manifestations

Primary pulmonary coccidioidomycosis

- About 60% of newly infected persons develop no symptoms following inhalation of arthrospores. The remainder develop symptoms after 1–4 weeks.
- Higher levels of exposure increase the likelihood of acute symptomatic disease.
- Most patients develop a mild or moderate flu-like illness that resolves without treatment.
- Up to 50% of patients develop a mild, diffuse, erythematous or maculopapular rash.
- Erythema nodosum or erythematous multiforme seen in 30% of cases, more common in women.
- Segmental pneumonia is most common radiological finding.
- About 20% of cases develop enlarged hilar lymph nodes or a pleural effusion.
- Single or multiple nodules, thick- or thin-walled cavities, as well as enlarged mediastinal lymph nodes can occur.

Chronic pulmonary coccidioidomycosis

- Small number of patients with primary disease are left with benign residual lesions.
- Most patients are asymptomatic but haemoptysis may occur in up to 25% of cases.
- Residual cavities can enlarge and rupture.
- In immunocompromised, acute pulmonary disease can be fatal.
- In immunocompetent, illness can mimic tuberculosis.

Disseminated coccidioidomycosis

- Fewer than 1% of patients develop disseminated disease.
- Progressive and fatal.
- Men five times more affected than women; ratio is reversed if women are pregnant.
- Disease develops within 12 months of initial infection.
- Disease can develop much later due to reactivation of quiescent lesions.
- Cutaneous, soft tissue, bone, joint and meningeal disease are most common.

• In immunosuppressed, widespread dissemination often occurs.
• Meningitis most serious complication; occurs in 30–50% of patients with disseminated disease; often results in hydrocephalus; fatal if not treated.

Coccidioidomycosis in AIDS

• Most cases recently acquired in endemic areas.
• Pulmonary disease most common presentation; chest radiographs show diffuse, reticulonodular infiltrates; more than 70% of patients die within 1 month despite treatment.

Essential investigations

Microscopy

Large, thick-walled endospore containing spherules can be seen on direct microscopy of pus, sputum or joint fluid but less commonly in blood.

Culture

C. immitis isolated from sputum, joint fluid, CSF sediment, pus and other specimens.

Identifiable mycelial colonies can be seen after incubation at 25–30°C for 2–7 days.

Skin tests

Skin tests do not distinguish between present or past infection.

Conversion from a negative to a positive result suggests recent infection.

False-negative results are common in anergic patients with disseminated disease.

Serology

Detection of specific IgM is useful in diagnosing acute infection; it appears within 4 weeks from onset of infection and disappears after 2–6 months.

IgM can be detected by latex agglutination (LA), test-tube precipitation (TP), or immunodiffusion (ID).

Specific IgG is useful for detecting later stages of coccidioidomycosis; titres rise with the progression of the disease.

IgG can be detected using complement fixation (CF) or ID. Rising CF titre is indicative of progressive disease.

Detection of antibodies in CSF is indicative of meningitis.

Management

Primary pulmonary coccidioidomycosis

This condition is normally self-limited and recovers without antifungal treatment.

However, a few patients require treatment to prevent progression.

Treatment is indicated in nonimmunocompromised patients with persistent symptoms, persistent debilitation, extensive or progressive pulmonary involvement, persistent hilar or mediastinal lymph node enlargement, rising or elevated antibody titre, or negative skin tests.

- Treatment of choice: amphotericin B 0.4–0.6 mg/kg/day.
- Once stabilized, 0.8–1.0 mg/kg every 48 hours. Achieve total dose of 0.5–1.5 g.

For milder infections: itraconazole or fluconazole 400 mg/day for 2–6 months.

Chronic pulmonary coccidioidomycosis

Enlarging cavities require surgical resection plus

- amphotericin B 0.4–0.6 mg/kg/day for 4 weeks, commencing 2 weeks before surgery.

For chronic progressive pneumonia:

- amphotericin B 0.4–0.6 mg/kg/day
- fluconazole 200–400 mg/day is an effective alternative, but has a high relapse rate.

Disseminated coccidioidomycosis

For nonmeningeal disease:

- Amphotericin B 1.0–1.5 mg/kg/day. Continue treatment until 2.5–3.0 g total dose reached.

However, there is poor response, with relapses common.

- Itraconazole 400 mg/day.
- Ketoconazole 400 mg/day.

- Fluconazole 400 mg/day effective in cutaneous, soft-tissue, bone or joint lesions.

Surgical debridement is necessary for osteomyelitis.

For AIDS patients:

- Amphotericin B 1.0–1.5 mg/kg/day. Continue until total dose of 1.0 g reached, then switch to itraconazole 400 mg/day or fluconazole 400 mg/day if there is improvement.

Meningitis

- Fluconazole 400 mg/day drug of choice, improvement in up to 80% of patients. Response rate in AIDS is lower.

However, fluconazole does not eradicate meningeal infection; the drug must be continued for life to prevent relapse.

Treatment with itraconazole 400 mg/day is under evaluation.

Histoplasmosis

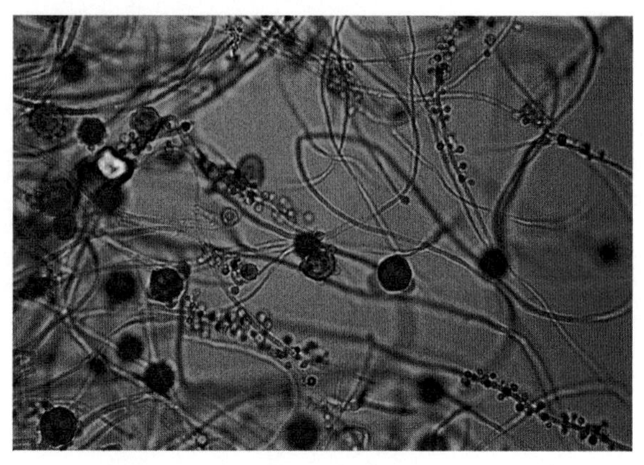

Histoplasma capsulatum conidia.

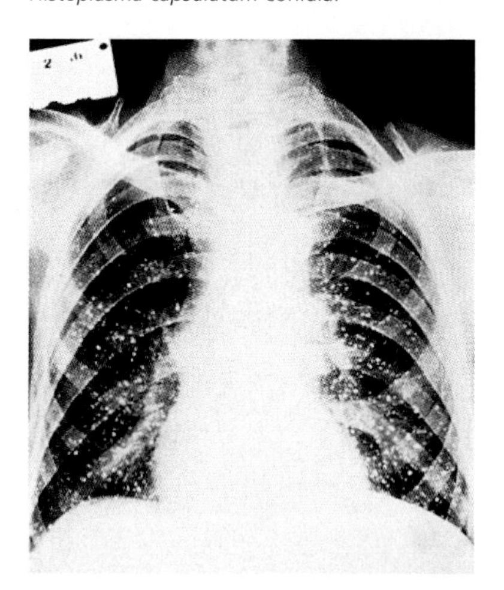

Radiographic appearance of chronic pulmonary histoplasmosis.

Definition

A mild and transient pulmonary infection in normal individuals caused by the dimorphic fungus *Histoplasma capsulatum*. Can proceed to a chronic infection of the lungs or more widespread infection in predisposed patients.

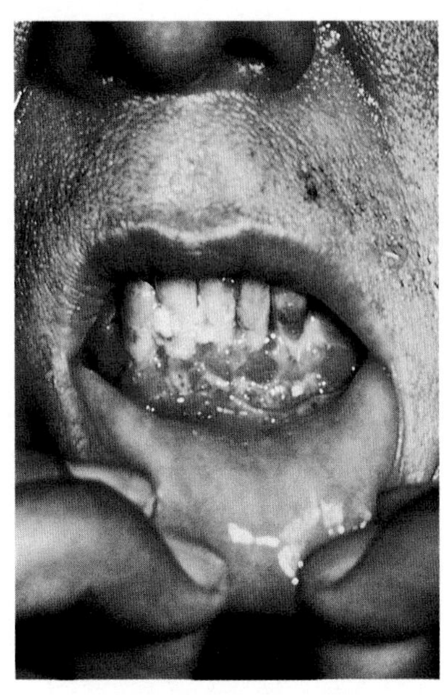

Mucosal ulcer in a patient with histoplasmosis.

Geographical distribution

Most prevalent in central North America, Central and South America. Other endemic areas include Africa, Australia, India and Malaysia.

Causal organism and habitat

- *H. capsulatum* exists as mycelium in nature, in tissue forms small round budding cells.
- Two varieties recognized: *capsulatum* and *duboisii*.
- Found in soil enriched with bird or bat droppings.
- Large numbers of spores dispersed into atmosphere.
- Large numbers of individuals can become infected.
- *H. capsulatum* var. *duboisii* confined to central Africa.

Clinical manifestations

Acute pulmonary histoplasmosis

- Normal individuals who inhale large numbers of spores develop acute symptomatic and often severe infection after 1–3 week incubation period.
- Nonspecific flu-like illness in symptomatic patients, resolves without treatment.
- Common symptoms include fever, chills, headache, myalgia, loss of appetite, cough and chest pain. In addition:
  - 10% of patients present with aseptic arthritis or arthralgia associated with erythema multiforme or nodosum
  - normal chest radiographs in most patients
  - hilar lymph node enlargement often evident
  - infiltrates heal over several month period to form a histoplasmoma, sometimes enlarges.
- Reinfection results in similar illness but distinct differences:
  - milder and occurs over much shorter incubation period
  - radiological signs different from newly infected individuals
  - no mediastinal lymph node enlargement, pleural effusions not seen.

Chronic pulmonary histoplasmosis

- Slowly progressive illness seen in middle-aged men with underlying chronic obstructive lung disease.
- A transient, segmental pneumonia that frequently progresses to fibrosis and cavitation with significant lung destruction.
- If left untreated, death can result from progressive lung failure.
- In patients with pneumonia, symptoms include a productive cough, fever, chills, weight loss, malaise, night sweats and pleuritic chest pain:
  - on radiography, interstitial infiltrates in apical segments of upper lung lobes.
- In patients with chronic fibrosis and cavitation there may be cough and sputum production. In addition:
  - haemoptysis in 30% of patients
  - fever, chest pain, fatigue and weight loss
  - radiographs reveal progressive cavitation and fibrosis

  - lesions more common in right upper lobe
  - pleural thickening adjacent to lesions found in 50% of patients.

Disseminated histoplasmosis

- Progressive, often fatal, associated with T-cell mediated immunological defects.
- Treatment essential.
- In infants and immunosuppressed, symptoms include high fever, chills, prostration, malaise, loss of appetite and weight loss. Also:
  - liver and spleen enlarged
  - liver function tests abnormal
  - anaemia common.
- In nonimmunosuppressed patients:
  - indolent, chronic course
  - hepatic infection common
  - adrenal gland destruction common
  - mucosal lesions seen in more than 60% of patients.
- Meningitis chronic complication:
  - 10–25% of patients with indolent disseminated infection
  - most patients abnormal CSF
  - *H. capsulatum* often isolated
  - occasional endocarditis and mucosal ulcerations in gastrointestinal tract.

African histoplasmosis

- Indolent in onset.
- Skin and bones predominant sites.
- Involvement of liver, spleen and other organs causes fatal wasting illness.
- Cutaneous lesions common.
- Both nodules and papules often enlarge and ulcerate.
- Osteomyelitis occurs in about 30% of patients:
  - spine, ribs, cranial bones, sternum and long bones most common sites
  - lesions often painless
  - spread into joints causes arthritis
  - spread into adjacent soft tissue causes purulent subcutaneous abscesses.

Histoplasmosis in AIDS patients

- AIDS defining illness – occurs in 2–5% of patients with AIDS.
- Acute infection or reactivation of an old latent infection.
- Most patients with AIDS present with disseminated histoplasmosis.
- Nonspecific symptoms, such as fever or weight loss.
- Up to 25% of patients have enlarged liver and spleen.
- Up to 25% of patients have anaemia, leucopenia and thrombocytopenia.

Essential investigations

Microscopy

All material must be examined as stained smears. Small oval budding cells are often seen within macrophages. There is a possibility of confusion with *Candida glabrata, Penicillium marneffei* and the small non-encapsulated cells of *Cryptococcus neoformans*.

Culture

Culture provides the definitive diagnosis, although unequivocal identification of culture requires conversion to yeast form, or exoantigen testing. The cultures should be incubated at 25–30°C for 4–6 weeks.

Serology

Immunodiffusion (ID) and complement fixation (CF) are positive in about 80% of patients. CF is more sensitive than ID, but ID is more specific.

False-negative reactions occur in immunosuppressed individuals with disseminated disease.

The detection of antigen in blood and urine is the most useful test for disseminated disease in AIDS.

Management

Acute pulmonary histoplasmosis

There is usually spontaneous improvement in this condition. However, treatment is indicated where there is no resolution of symptoms after 2–3 weeks. In this case, treat with:

- amphotericin B 0.5-0.7 mg/kg/day for 2–4 weeks; or
- oral itraconazole 400 mg/day for 6 months.

Chronic pulmonary histoplasmosis

If cavitation is not present and the symptoms are mild, delay treatment if healing is apparent. In cases where the symptoms progress, treat with:

- oral itraconazole 400 mg/day for 6 months.

If itraconazole is contraindicated, treat with:

- amphotericin B 0.5-0.7 mg/kg/day for 10 weeks.

Follow up patients for at least 12 months after discontinuation of treatment.

Disseminated histoplasmosis

For nonimmunosuppressed patients, treat with:

- oral itraconazole 400 mg/day; or
- amphotericin B 0.5-0.7 mg/kg/day for 10 weeks; but for infants 1.0 mg/kg for at least 6 weeks.

For AIDS patients and patients with African histoplasmosis, treat with:

- amphotericin B 0.5-0.7 mg/kg/day; or
- intravenous itraconazole 400 mg/day .

Relapse is common; therefore a maintenance treatment should be given:

- amphotericin B 1.0 mg/kg dose at 1- or 2-week intervals.

For mild forms: the intravenous itraconazole can be followed by the oral solution of itraconazole. A maintenance treatment with fluconazole 400 mg/day is suitable if itraconazole is contraindicated.

Paracoccidioidomycosis

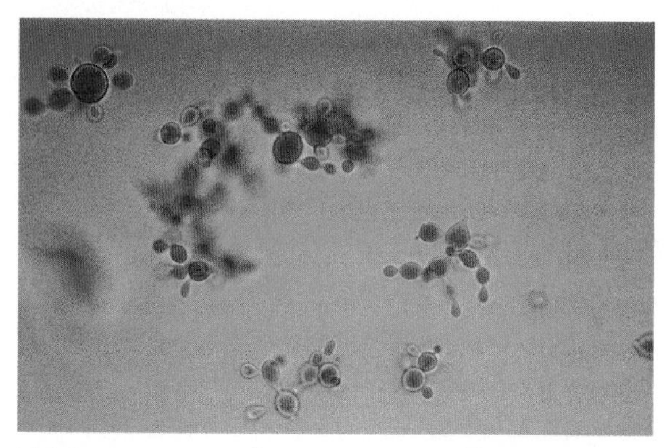

Microscopic appearance of *Paracoccidioides brasiliensis*.

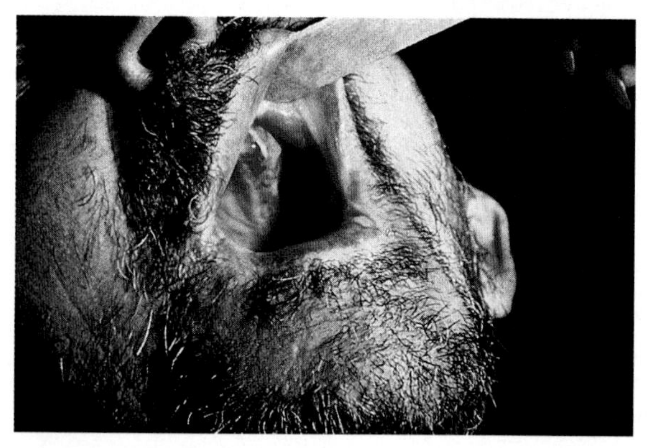

Oral lesions in a patient with paracoccidioidomycosis.

Definition

A benign and transient pulmonary infection caused by *Paracoccidioides brasiliensis*.

Later reactivation results in chronic infection of the lungs or other organs, in particular the skin and mucous membranes.

Geographical distribution

Most cases are in South and Central America.

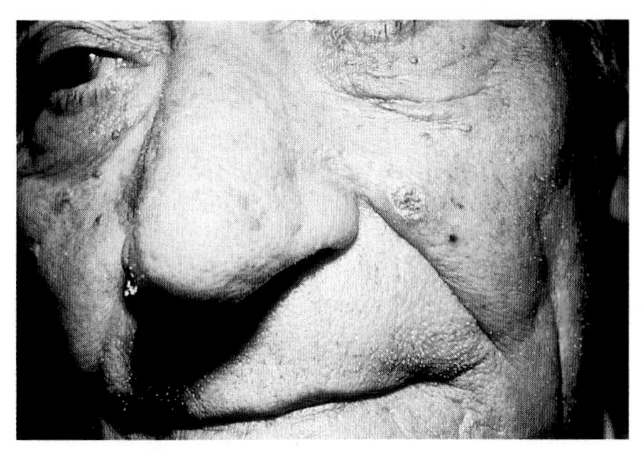

Cutaneous lesions in a patient with paracoccidioidomycosis.

Infection occurs among individuals who have visited an endemic region.

Causal organism and habitat

- *P. brasiliensis* is dimorphic; it grows in nature as mycelium.
- Large oval or globose cells with characteristic multiple buds encircling the mother cell in tissue.

Clinical manifestations

Chronic pulmonary paracoccidioidomycosis

- Symptomatic disease most prevalent between ages of 30 and 50 years. Most cases in men – associated with outdoor occupations.
- Lungs are normal site of infection and symptoms include productive cough, fever, night sweats, malaise, haemoptysis and weight loss.
- Spread through lymphatics to regional lymph nodes and then mucosa, skin and other organs.
- Normal presentation: chronic progressive infection as a result of reactivation of old quiescent lesions.
- Indolent in onset, appearing long after an individual has left an endemic region.
- Characteristic radiological findings, but not diagnostic.
- Multiple bilateral interstitial infiltrates often found.

- Hilar lymph node enlargement found in 50% of cases.
- Lesions must be distinguished from histoplasmosis and tuberculosis.

Mucocutaneous paracoccidioidomycosis

- Ulcerative mucocutaneous lesions are most obvious presentation.
- Mouth and nose most common sites.
- Perforation of palate or nasal septum may occur.
- Lymphadenopathies common in patients with buccal cavities.

Disseminated paracoccidioidomycosis

- Haematogenous and lymphatic spread can lead to widespread infection.
- Nodular or ulcerated lesions of small or large intestine, hepatic and splenic lesions, adrenal gland destruction.

Essential investigations

Microscopy

Characteristic large, multiple budding cells are seen in pus, sputum and crusts from granulomatous lesions.

Culture

Culture provides the definitive diagnosis. The cultures should be incubated at 25–30°C for 2–3 weeks and thereafter retained for 4 weeks.

Subculture on blood agar at 37°C will induce sporulation.

Serology

Complement fixation is positive in more than 90% of cases.

There is a cross-reaction with blastomycosis, histoplasmosis and sporotrichosis.

Immunodiffusion is positive in 80–90% of cases with the active disease.

Titres decline following successful treatment.

Management

Long-term treatment is needed and relapse is common. Treat with:

- oral itraconazole 100 mg/day for 6 months
- ketoconazole 200–400 mg/day for up to 12 months is an alternative, but less well tolerated.

If itraconazole and ketoconazole are contraindicated fluconazole 200–400 mg/day for 6 months is also effective.

Chromoblastomycosis

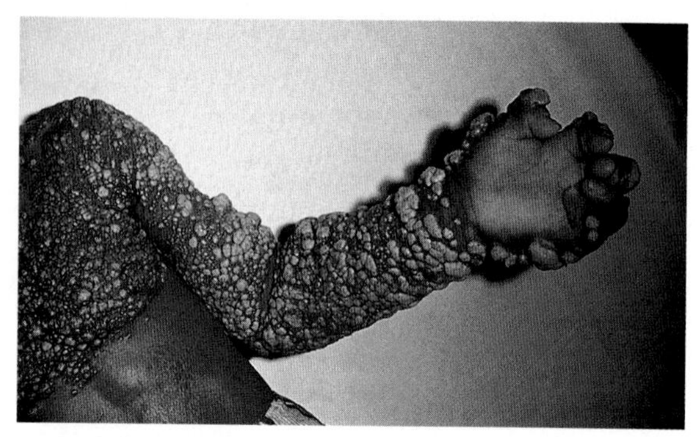

Chromoblastomycosis.

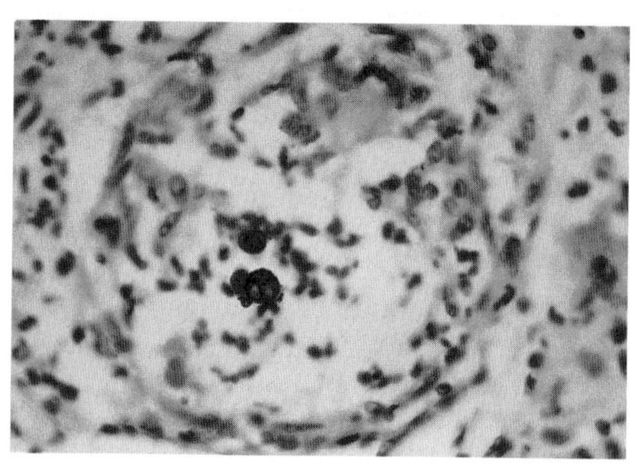

Histopathological appearance of chromoblastomycosis.

Definition

Chromoblastomycosis (chromomycosis) is a chronic, localized infection of the skin and subcutaneous tissue of the limbs and characterized by raised, crusted lesions.

Geographical distribution

Most common in tropical and subtropical regions with most cases occurring in South and Central America.

Causal organisms and habitat

- Caused by brown-pigmented (dematiaceous) fungi such as *Phialophora verrucosa*, *Fonsecaea pedrosoi*, *Fonsecaea compacta*, *Cladophialophora (Cladosporium) carrionii* and *Rhinocladiella aquaspersa*.
- Widespread in the environment in soil, wood and plant material.
- Infection follows traumatic inoculation of fungus into skin.
- Disease is most prevalent among individuals with outdoor occupations.

Clinical manifestations

- Common sites: lower legs, feet, hands, arms, back and neck.
- Initial lesion is a small, pink, painless papule. However, lesion increases in size if left untreated and becomes a large hyperkeratotic plaque.
- Lymphatic spread with satellite lesions around original lesion.
- Advanced disease:
  - some lesions become pedunculated
  - bacterial superinfection
  - lymph stasis and elephantiasis in some patients.

Essential investigations

Microscopy

Microscopy reveals clusters of small, round, thick-walled, brown sclerotic cells in tissue sections or wet preparations of pus, scrapings or biopsies.

Culture

Culture provides a definitive diagnosis. Olive green–brownish black mycelial colonies appear after 1–2 weeks at 25–30°C. Cultures should be retained for four weeks before discarding.

Management

This condition is difficult to treat

- Oral itraconazole 200–600 mg/day for 12–36 months
- Local application of heat is beneficial.

Subcutaneous zygomycosis

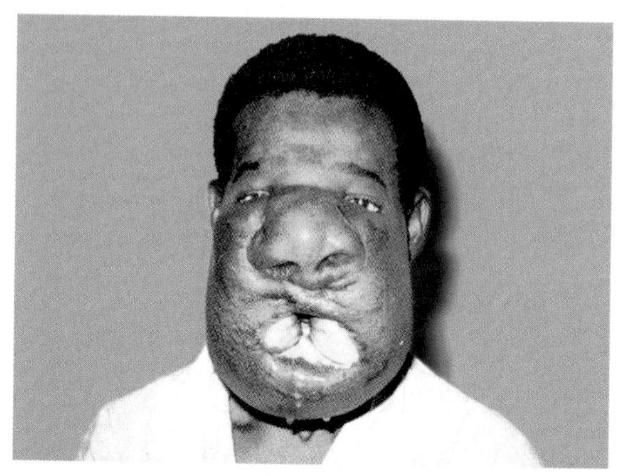

Rhinofacial conidiobolomycosis.

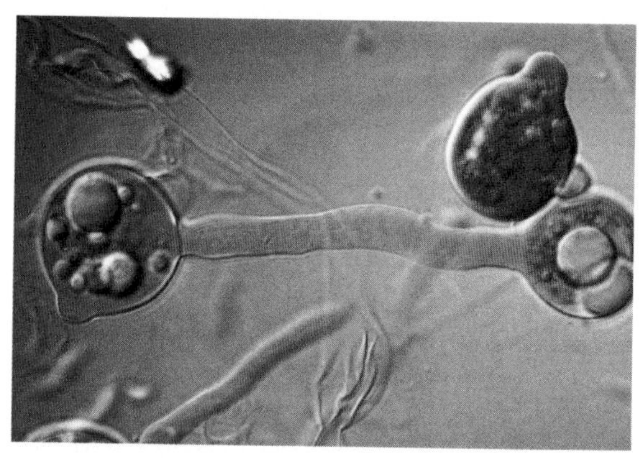

Microscopic appearance of *Conidiobolus coronatus* spores.

RHINOFACIAL CONIDIOBOLOMYCOSIS

Definition

Chronic localized subcutaneous fungal infection. Originates in nasal mucosa and spreads to adjacent subcutaneous tissue of the face causing severe disfigurement.

Geographical distribution

It is most common in West Africa, in particular Nigeria.

Causal organism and habitat

- *Conidiobolus coronatus* (*Entomophthora coronata*).
- Found in soil and decomposing vegetation in tropical rain forests.
- Insect pathogen.

Clinical manifestations

- More common in men than in women or children.
- Associated with living or working in tropical rain forests.
- Infection originates in nasal mucosa – inferior turbinates.
- Autoinoculation from soiled hands.
- Common symptoms include nasal obstruction, often unilateral, and nasal discharge.
- Progression of disease is slow but relentless.
- Tissue swelling becomes pronounced – affects forehead, nose, cheeks, upper lip.
- No bone involvement.
- Lesions have distinct margin.
- Skin stretched but not broken.

Essential investigations

Microscopy

Broad, nonseptate, thin-walled mycelial fragments are seen in mucosal smears.

Culture

Culture is difficult to obtain. Use a wide range of media. Incubate at 25–35°C.

Management

This is difficult to treat.

- Good response to oral itraconazole 200–400 mg/day.
- Ketoconazole 200–400 mg/day.

Continue treatment for at least 1 month after clearance of lesions.

- Saturated potassium iodide 1 ml three times daily, increase up to 4–6 ml three times daily. Continue for at least 1 month after resolution of lesions.

BASIDIOBOLOMYCOSIS

Definition

This is a chronic subcutaneous infection of trunk and limbs.

Geographical distribution

Most common in East and West Africa.

Causal organism and habitat

- *Basidiobolus ranarum* (*B. meristosporus*, *B. haptosporus*).
- Decomposing vegetation lying on soil.
- Intestines of frogs, toads, lizards and other small reptiles.

Clinical manifestations

- More common in children and adolescents than in adults.
- Transmission probably by traumatic inoculation.
- Common sites: buttocks, thighs, arms, legs and neck.
- Disfiguring.
- Lymphatic obstruction can result in elephantiasis.

Essential investigations

Microscopy

In histopathological sections wide, irregular, nonseptate filaments or hyphal fragments can be seen.

Culture

Culture specimens on glucose peptone agar at 30°C. Identifiable colonies should be obtained in less than 1 week.

Management

Topical application of saturated potassium iodide solution is a possible treatment. However,

- cotrimoxazole is sometimes more effective: two tablets, three times daily.

In either case, continue treatment for at least 1 month after resolution of the lesions.

- Oral ketoconazole 400 mg/day is sometimes useful.

Lobomycosis

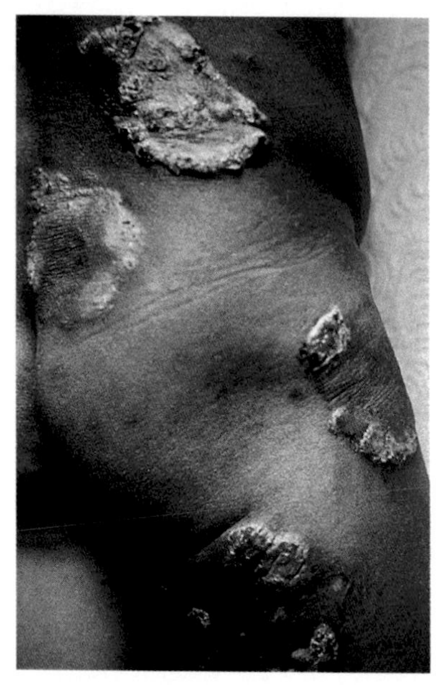

Lobomycosis.

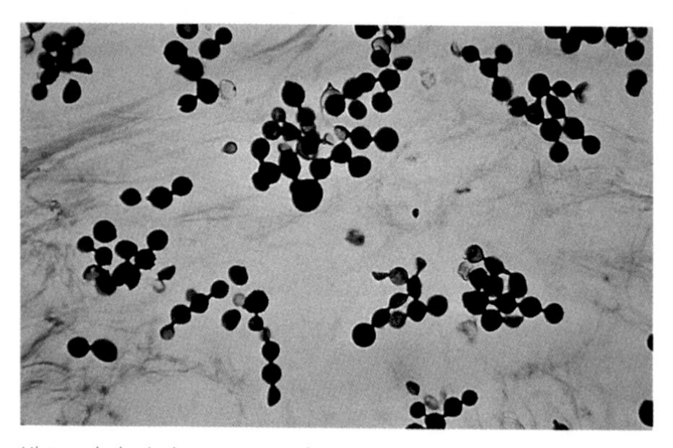

Histopathological appearance of *Loboa loboi* in tissue.

Definition

Lobomycosis (keloidal blastomycosis) is a rare, chronic

infection of skin and subcutaneous tissue due to *Loboa loboi*.

Geographical distribution

Most cases occur in the Amazon region of central Brazil and in Surinam.

Causal organism and habitat

- *Loboa loboi*.
- All attempts to isolate in culture have failed.
- Round or elliptical cells in tissue.
- Natural habitat unknown.
- Water has an important role in infection.
- Infection related to a traumatic incident.
- Disease more common in men than in women or children.
- Most prevalent 30–40 years of age.

Clinical manifestations

- Most common sites: legs, arms, face, ears, buttocks.
- Initial lesion: papule or a small nodule, slowly proliferates to form extensive keloidal or verrucous lesions in dermis.
- Autoinoculation leads to further lesions.
- Disease symptomless in most cases.

Essential investigations

Microscopy

Microscopy reveals large numbers of large, round or oval, thick-walled cells (over 10 μm in diameter). The cells produce multiple buds that resemble the tissue form of *Paracoccidioides brasiliensis*. The cells often form in unbranched chains.

Management

Antifungal treatment is ineffective. However, oral clofazimine has given promising results. The treatment of choice is surgical excision if the lesions are not too extensive.

Mycetoma

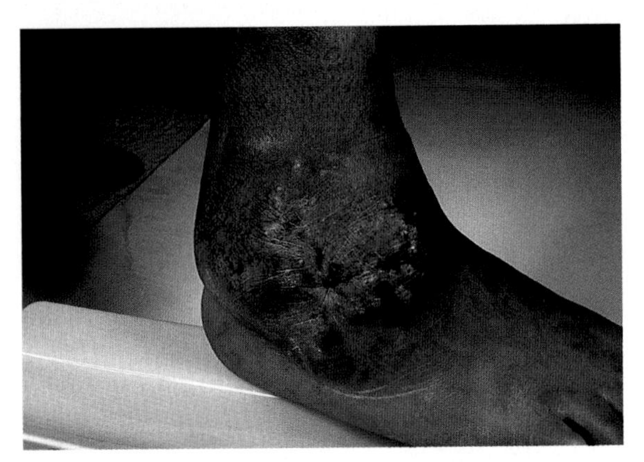

Mycetoma of foot.

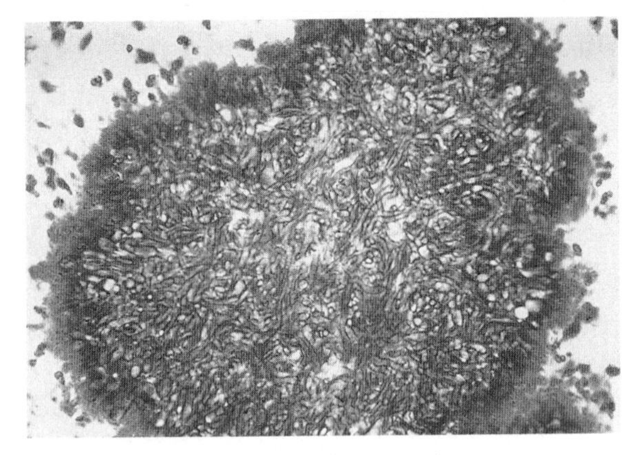

Microscopic appearance of *Madurella grisea* grain.

Definition

Chronic localized destructive infection of the skin, subcutaneous tissue and bone of feet or hands.

May be caused by various species of fungi (eumycetoma) or actinomycetes (actinomycetoma).

Geographical distribution

It is found in arid tropical and subtropical regions of Africa and Central and South America. It is endemic in countries surrounding the Saharan and Arabian deserts, India and the Far East.

Causal organisms

- Fungi and actinomycetes, isolated from soil and vegetation.
- Three species of fungi are common causes of eumycetoma: *Madurella mycetomatis*, *M. grisea* and *Scedosporium apiospermum*.
- Three species of actinomycetes are associated with most cases of actinomycetoma: *Actinomadura madurae*, *Streptomyces somaliensis* and *Nocardia brasiliensis*.

Clinical manifestations

- Infection follows traumatic implantation into skin or subcutaneous tissue.
- Mycetomas most common on feet (70% of cases), followed by hands (10% of cases). Other sites include back, neck and back of head.
- Clinical features of disease are similar, regardless of organism.
- Initial lesions are small, firm painless nodules.
- Eumycetomas follow a slower, less destructive course than actinomycetomas:
  - eumycetomas remain localized – swelling and destruction of adjacent anatomical structures occurs late in course of disease
  - actinomycetomas have less well-defined margins, merge with surrounding tissue. Progression is more rapid. Involvement of bone is earlier and more extensive.
- Lesions present as swellings covered with hypo- or hyperpigmented skin.
- Lesions develop single or multiple sinus tracts.
- Sinus tracts discharge pus containing grains onto skin surface.
- Infection spreads to adjacent tissue, including bone.

- Movement of joints may be impaired.
- Radiology useful in determining extent of bone involvement.
- Cavities in bone vary in size.
- Spread to adjacent tissue is common.
- Bacterial superinfection is common.

Essential investigations

Microscopy

Direct microscopy will confirm diagnosis. Actinomycotic grains contain very fine filaments whereas fungal grains contain short hyphae, which are sometimes pigmented.

- black grains – suggests a fungal infection
- minute white grains – nocardia
- large white grains – fungal or actinomycotic
- small red grains – actinomycotic or fungal
- yellowish-white grains – actinomycotic or fungal.

Culture

Culture grains on glucose peptone agar at 25–30°C and at 37°C for up to 6 weeks.

Management

It is essential to distinguish eumycetoma from actinomycetoma.

In cases of actinomycetoma:

- streptomycin sulphate 1000 mg/day intramuscular injection
- streptomycin and co-trimoxazole, two tablets in the morning and two in the evening
- infection due to *A. madurae* and *S. somaliensis*, combine streptomycin with dapsone, 200 mg/day.

If there is no response:

- streptomycin plus rifampicin 600 mg/day
- streptomycin plus sulphadoxine–pyrimethane in the form of one Fansidar tablet twice a week. Each tablet contains 500 mg sulphadoxine and 25 mg pyrimethamine.

In cases of eumycetoma:

- unresponsive to antifungals
- radical surgical removal is best option

- amputation if bone involvement
- relapse common if not all infected tissue removed
- long-term treatment with ketoconazole 400 mg/day or itraconazole 400 mg/day successful in some cases.

Rhinosporidiosis

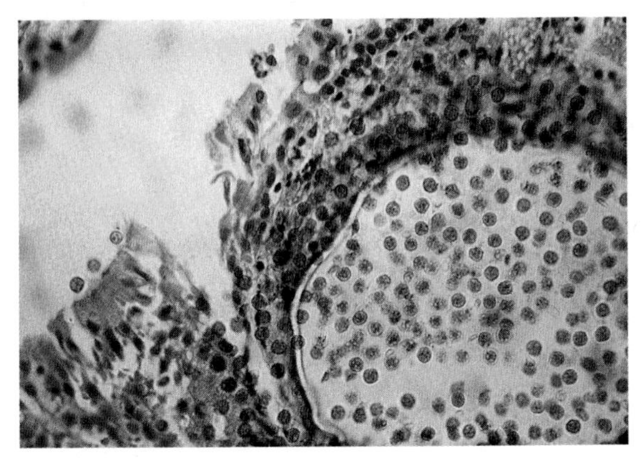

Histopathological appearance of rhinosporidiosis showing thick-walled sporangium.

Definition

Granulomatous infection of the nasal and other mucosa.

Geographical distribution

Most common in India and Sri Lanka. However, there have been sporadic cases in East Africa, Central America and southeast Asia.

Causal organism and habitat

- *Rhinosporidium seeberi.*
- Attempts to isolate in culture have failed.
- In tissue, forms abundant large thick-walled sporangia.
- Large numbers of endospores produced within sporangium.
- Not known how infection is acquired, however, pools of stagnant water may be an important source.
- Disease most prevalent in rural districts.
- Most common in children and adolescents, with males more commonly affected than females.

Clinical manifestations

- Nose most common site, with large sessile or pedunculated lesions in one or both nostrils.
- Insidious infection.
- Rhinoscopic examination reveals papular or nodular, smooth-surfaced pink, red or purple lesions that become pedunculated.
- If located low in nostril, polyps may protrude and hang onto the upper lip.
- General health is not usually impaired.
- If left untreated, polyp will continue to enlarge.

Essential investigations

Microscopy

Microscopy of tissue sections or wet preparations of tissue or discharge show large round or oval sporangia up to 350 μm in diameter. The sporangium may be filled with endospores.

R. seeberi has never been isolated in culture.

Management

Surgical excision of the lesions is treatment of choice, although drug treatment has been effective. Recurrence is common.

Sporotrichosis

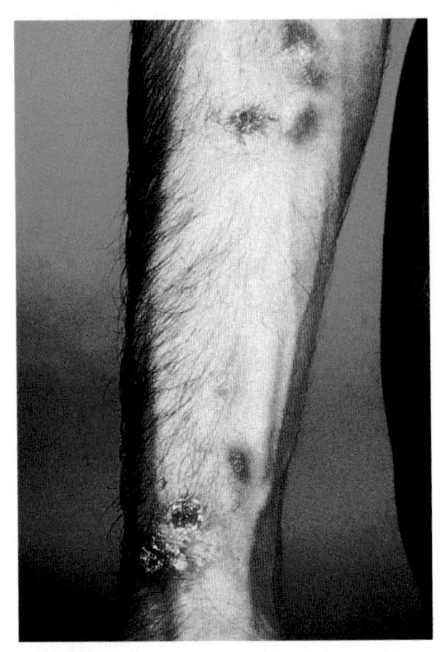

Sporotrichosis.

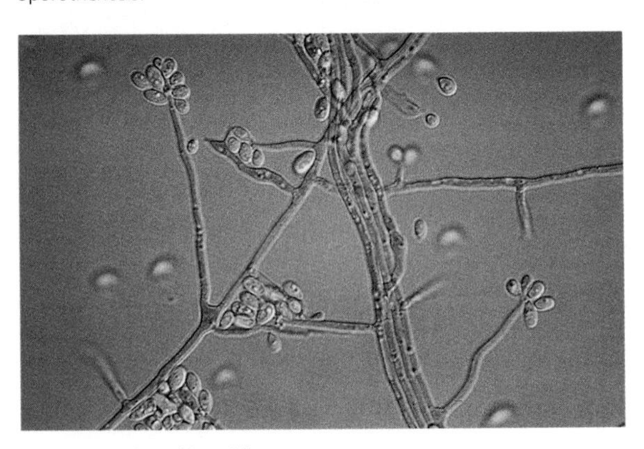

Sporothrix schenckii conidia.

Definition

Subacute or chronic cutaneous or subcutaneous infection caused by *Sporothrix schenckii*. Commonly shows lymphatic

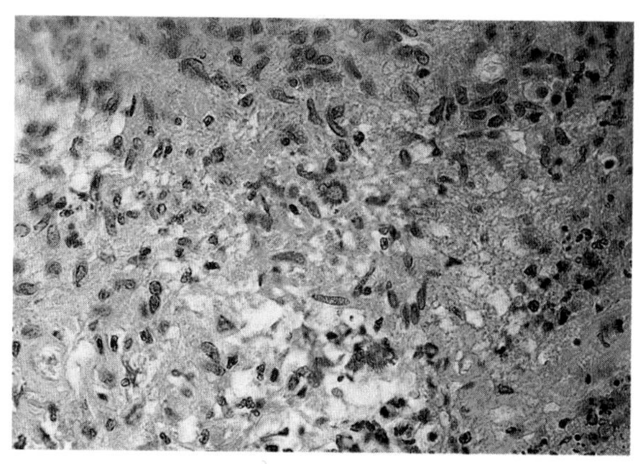

Histopathological appearance of sporotrichosis showing asteroid bodies.

spread. Occasionally, infection of lungs, joints and bones occurs in predisposed individuals.

Geographical distribution

World-wide, although it is most common in warm, temperate or tropical climatic regions. Most cases are from the USA although it is also endemic in Central and South America, Africa and Australia.

Causal organism and habitat

- *Sporothrix schenckii.*
- Soil, on plants, on plant materials.
- Mycelium in nature, small budding cells in tissue.
- More common in adults and most prevalent where there is contact with soil, plants, or plant materials.
- Most cases sporadic.
- Epidemics sometimes occur in endemic regions.

Clinical manifestations

Cutaneous sporotrichosis

- Follows traumatic implantation of fungus into skin or subcutaneous tissue.

- Minor trauma sufficient to introduce organism.
- Affects exposed sites, in particular hands and fingers.
- Initial lesion appears 1–4 weeks after traumatic incident as a small, firm, painless nodule. After this:
  - skin becomes violaceous
  - nodule becomes soft to form persistent discharging ulcer with irregular edge
  - ulcer becomes oedematous and crusted
  - further nodules develop along course of lymphatic channels, and become ulcerated.
- No lymphatic spread in about 25% of cutaneous infections.
- Disseminated cutaneous forms occur occasionally.

Extracutaneous sporotrichosis

- Most commonly where there is underlying disease.
- Most common sites: lungs, joints and bones – resulting in arthritis.
- Pulmonary disease uncommon.

Essential investigations

Microscopy

The organism is seldom seen in pus or tissue. Detection of oval shaped cells or asteroid bodies confirms diagnosis.

Culture

Culture provides the definitive diagnosis. Use several media, including glucose peptone agar. On this, mycelial colonies appear in 3–5 days at 25–30°C.

Confirmation of identification requires conversion to yeast form on blood agar at 37°C.

Management

Cutaneous and lymphocutaneous forms

Treat with:

- oral itraconazole 100–200 mg/day for 3–6 months. Continue treatment for several months after the lesions have cleared.
- Alternatively, give fluconazole 400 mg/day minimum if itraconazole is not tolerated or not absorbed.

Saturated potassium iodide is useful where antifungals are not available.

Local application of heat may be used in cases of drug intolerance.

Extracutaneous forms

These are difficult to treat. For osteoarticular forms potassium iodide is ineffective. Therefore:

- itraconazole 400 mg/day for more than 12 months is the drug of choice
- alternatively, if itraconazole is contraindicated, fluconazole 400–800 mg/day can be given.

For pulmonary forms:

- amphotericin B 1.0 mg/kg/day. When there are signs of improvement, substitute with itraconazole 400 mg/day.

For disseminated forms:

- amphotericin B 1.0 mg/kg/day until 1–2 g have been administered. Less acute forms can be treated with itraconazole 400 mg/day
- in AIDS itraconazole treatment must be maintained for life.

Hyalohyphomycosis

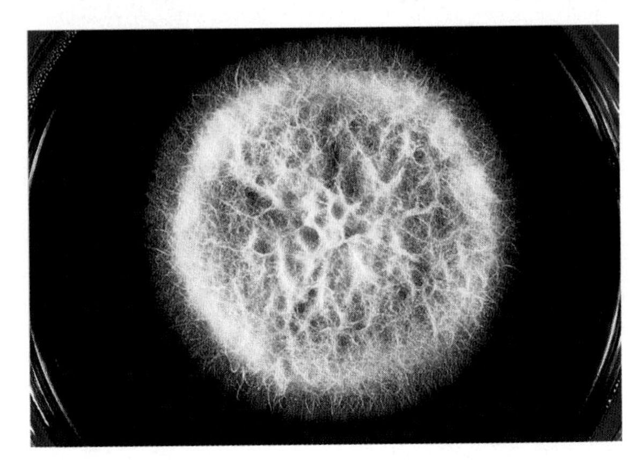

Fusarium solani culture.

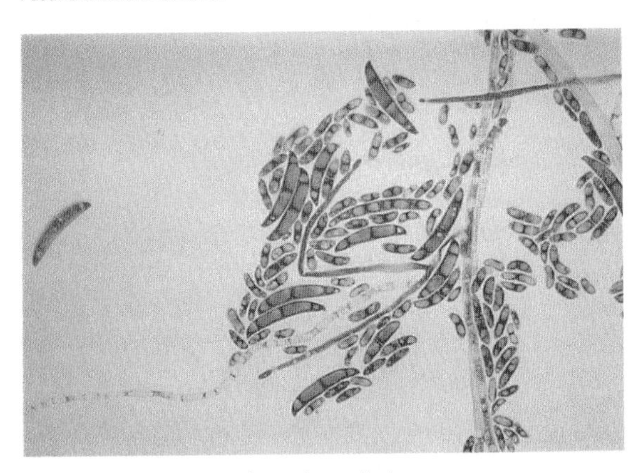

Microscopic appearance of *Fusarium solani*.

Definition

Hyalohyphomycosis is the term used to describe infections with moulds that appear as hyaline (colourless), septate filaments in host tissues.

Geographical distribution

World-wide.

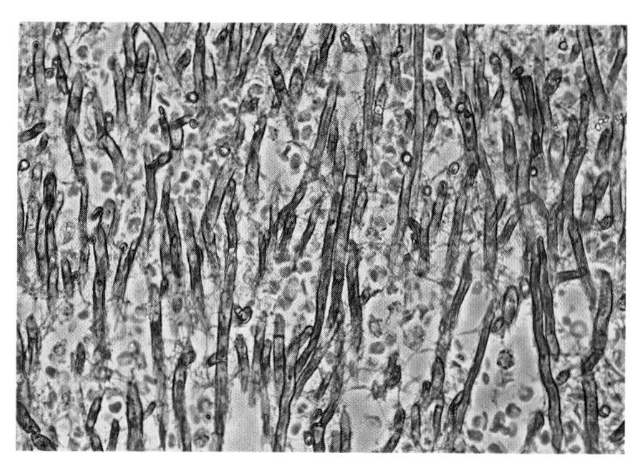

Histopathological appearance of *Scedosporium apiospermum* in a heart valve.

Causal organisms and habitat

- More than 40 species from more than 20 genera implicated.
- Infections caused by environmental moulds that produce large numbers of spores suited to air or water dispersal.
- Diseases sufficiently common to warrant own disease name are not included in this group. Best known example of hyalohyphomycosis is aspergillosis.
- Two most commonly isolated agents: *Fusarium* spp. and *Scedosporium* spp.
- Infections with *Scedosporium apiospermum* sometimes referred to as pseudallescheriosis after the sexual form of the fungus *Pseudallescheria boydii.*
- *Scedosporium prolificans* emerging as a pathogen in immunocompromised patients.
- Other emerging pathogens: *Acremonium* spp., *Paecilomyces* spp., *Scopulariopsis* spp. with many others isolated rarely.

Clinical manifestations

- Manifestations of hyalohyphomycosis often mimic those of invasive aspergillosis.
- Infections most commonly encountered in neutropenic patients.

- Neutropenic fever unresponsive to broad-spectrum antibiotics – most common presenting feature.
- Skin lesions may be present in up to 70% of cases of *Fusarium* infection.
- Inhalation thought to be the major route of acquisition.
- Vascular invasion may result in thrombosis and tissue necrosis.
- Infection with *Fusarium* spp. or *Scopulariopsis* spp. may arise from an infected nail in neutropenic patients.
- Infection with *Scedosporium* spp. may follow aspiration of contaminated water during near-drowning accidents.
- Dissemination to the brain is a common consequence of *Scedosporium* pneumonia.
- Fungus ball formation may be caused by *Scedosporium* in patients with pre-existing lung cavities.

Essential investigations

Microscopy

On microscopy, the hyphal tissue form cannot be distinguished from other agents of hyalohyphomycosis or aspergillosis.

Culture

The infecting organism may be isolated from biopsies of cutaneous lesions.

Culture is essential to make a definitive diagnosis. Blood culture is frequently positive in cases of *Fusarium* infection.

Management

Many of the agents of hyalohyphomycosis are refractory to treatment with currently available agents.

Aggressive surgical debridement of infected tissue is the treatment of choice in patients with localized lesions. However, the outcome is often dependent on the immune status of the host.

Cytokine therapy may be used in addition to antifungal therapy in an attempt to reduce the depth and duration of neutropenia.

Treatment with high dose amphotericin B may be effective in patients with *Fusarium* infection, use of a lipid preparation should be considered.

Anecdotal evidence and *in vitro* susceptibility testing suggest that the new triazole, voriconazole, may be of benefit in cases of *Fusarium* infection.

Scedosporium spp. are often resistant to amphotericin B *in vitro* and fail to respond to the drug *in vivo*.

Azole therapy with itraconazole 400 mg/day or voriconazole may be beneficial in cases of *Scedosporium* infection and there is anecdotal and laboratory evidence that an azole combined with terbinafine may have more activity than either drug alone.

Penicillium marneffei infection

Penicillium marneffei culture.

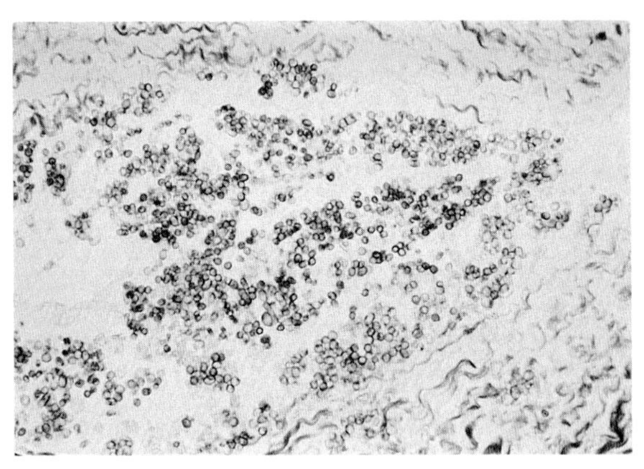

Histopathological appearance of *Penicillium marneffei* in the liver of an AIDS patient.

Definition

Penicillium marneffei is one of the most frequent opportunistic infection encountered in patients with AIDS who have resided in or visited southeast Asia or southern China. However, the disease can also occur in otherwise healthy patients.

Geographical distribution

Northern Thailand and southern China.

Causal organism and habitat

- Dimorphic fungus: mycelium at 37°C in tissue and at 28°C forms round to elliptical cells which divide by fission.
- Natural habitat has not been identified; probably soil.
- Third most common opportunistic infection among AIDS patients in northern Thailand.
- Increasingly diagnosed in patients with AIDS who have visited endemic areas.

Clinical manifestations

- Infection follows inhalation.
- Most patients present with widespread dissemination.
- Often chronic progressive illness.
- Most common symptoms are fever, weight loss, debilitation, multiple papular skin lesions, generalized lymphadenitis and hepatosplenic enlargement. It is fatal if left untreated.

Essential investigations

Microscopy and culture

Round, oval or elliptical cells, often with prominent cross-walls seen in Wright-stained bone marrow smears or touch smears of skin lymph node biopsies.

- Isolated from skin and lymph node biopsies, pus, bone marrow aspirates, sputum and BAL.
- Recovered from blood cultures in > 70% of AIDS cases.
- Mycelial cultures after 1 week at 25–30°C on glucose peptone agar; distinctive red pigmentation, hazard cat III.

Management

- Amphotericin B 1.0 mg/kg/day for 2 weeks
- Switch to itraconazole 200–400 mg/day for a further 6 weeks if patients show improvement.

For mild infections itraconazole can be used from the outset.

Maintenance with oral itraconazole 200 mg/day in AIDS patients; relapse is common when discontinued.

Phaeohyphomycosis

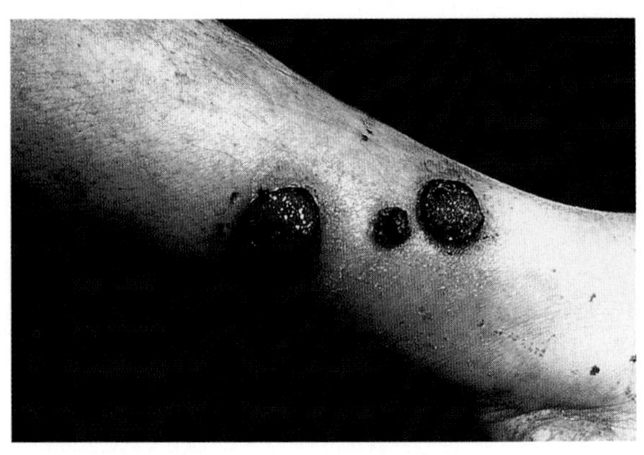

Cutaneous phaeohyphomycosis caused by an *Alternaria* spp.

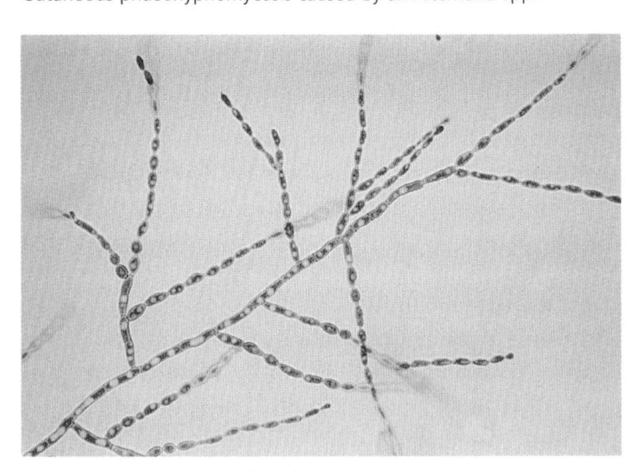

Microscopic appearance of *Cladophialophora bantiana*.

Definition

Phaeohyphomycosis is the term used to describe infections with brown-pigmented moulds.that appear as septate filaments in host tissues.

Geographical distribution

World-wide, but invasive forms have been reported more

frequently in temperate climates. Subcutaneous lesions are more common in tropical regions.

Causal organisms and habitat

• Agents of phaeohyphomycosis share common feature of melanin in cell wall producing a dark colour and possibly making them more resistant to damage by host phagocytic cells.
• More than 100 different species from at least 60 genera implicated and list increasing steadily.
• Infections caused by environmental moulds that produce large numbers of spores suited to air or water dispersal.
• Infection often follows inhalation or traumatic implantation.
• Recognized aetiological agents include *Alternaria* spp., *Bipolaris* spp., *Cladophialophora bantiana, Curvularia* spp., *Exerohilum* spp., *Exophiala* spp., *Phialophora* spp. and *Ochroconis* spp.
• *Exophiala jeanselmei* and *Exophiala* (*Wangiella*) *dermatitidis* implicated most frequently.
• *Cladophialophora bantiana* and *Exophiala dermatitidis* appear to be particularly neurotrophic.

Clinical manifestations

There are several different presentations depending on the route of acquisition.

The most common lesions are localized cutaneous or subcutaneous abscesses, granulomas or cysts that occur following traumatic implantation.

Such lesions should not be confused with those of chromoblastomycosis (see p. 75) in which muriform cells are seen or black-grain eumycetoma (see p. 84), which is characterized by the formation of dark granules.

Cutaneous or subcutaneous lesions of phaeohyphomycosis often follow minor trauma such as cuts, abrasions or splinters and there is little tendency for lymphatic or haematogenous dissemination. Draining sinus formation may occur in immunocompromised patients.

Paranasal sinus infection may follow inhalation.

In immunocompromised patients black, necrotic lesions may be visible.

Cerebral infection may progress from paranasal sinus infection or may be due to haematogenous dissemination from the lungs.

Infection of the brain is often indolent in onset.

Infection of other deep sites including the heart may follow surgical procedures.

Essential investigations

Microscopy

Microscopy of exudates or tissue sections reveals dark coloured, branching, septate hyphae.

Culture

Culture is essential to isolate and identify the aetiological agent and in order to perform antifungal susceptibility testing.

Prolonged incubation for up to 3 weeks at 30°C on Sabouraud dextrose agar may be required.

CT scans are invaluable in determining the extent of paranasal sinus infection and determining whether brain lesions are solitary or multifocal.

Management

Surgical excision is often curative in cases of localized infection whereas antifungal therapy alone with amphotericin B has often resulted in relapse.

Repeated surgical debridement is also an important adjunct to antifungal therapy in cases of paranasal sinus infection and cerebral phaeohyphomycosis.

Disseminated infections are asociated with a high rate of mortality especially if there are multifocal brain lesions.

Amphotericin B therapy possibly with the addition of flucytosine is indicated.

There is some evidence that itraconazole 100–400 mg/day has proved effective in cases of paranasal sinus infection.

Uncommon yeast infections

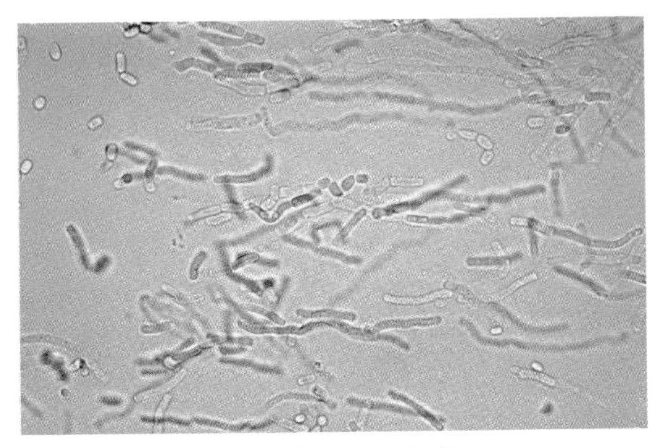

Microscopic appearance of *Trichosporon beigellii*.

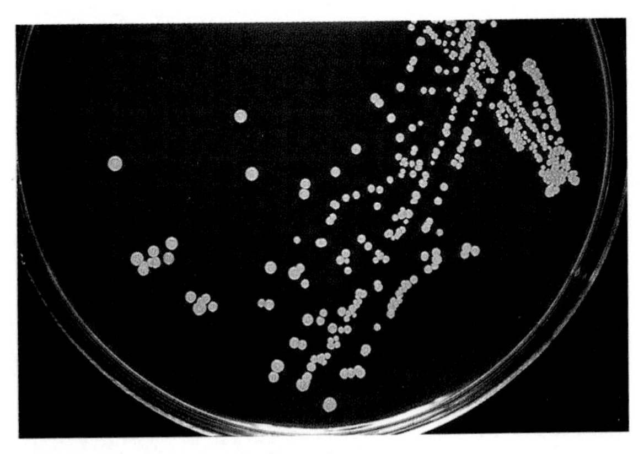

Culture of *Malassezia pachydermatis*.

TRICHOSPORONOSIS

Definition

A deep-seated infection in immunocompromised hosts caused by *Trichosporon beigelii (T. cutaneum)*. Similar infection caused by *T. capitatum* (reclassified as *Blastoschizomyces capitatus*).

Geographical distribution

World-wide.

Causal organisms and habitat

- *Trichosporon* spp., *Blastoschizomyces capitatus*.
- Soil, water and on plants.
- Mucosal and cutaneous surfaces.
- Endogenous reservoir in gastrointestinal tract.

Clinical manifestations

- Localized deep infections:
  - endophthalmitis
  - endocarditis
  - peritonitis
  - pulmonary.
- Disseminated:
  - uncommon, seen in neutropenia, BMT recipients, solid organ recipients and AIDS
  - many similarities with systemic candidosis.

Essential investigations

Microscopy

Microscopy will reveal branching hyphae, rectangular arthrospores and budding blastospores.

Culture

Culture of blood, urine and cutaneous lesions yields white to cream, heaped colonies.

Blood cultures and cutaneous lesion biopsies are often positive.

Serology

There is antigenic cross-reactivity with *Cryptococcus neoformans*.

Latex agglutination test for cryptococcosis positive in trichosporonosis.

Management

For localized infection in non-neutropenics treat with

- amphotericin B 1.0 mg/kg/day.

In neutropenics amphotericin B is of little benefit. Individual cases can be treated with fluconazole 400 mg/day.

SYSTEMIC *MALASSEZIA* (*PITYROSPORUM*) INFECTION

Definition

This is a serious systemic infection seen in low birth-weight infants, and debilitated adults and children receiving parenteral lipid nutrition through indwelling catheters.

Geographical distribution

World-wide.

Causal organism and habitat

- *Malassezia furfur* (*Pityrosporum orbiculare*, *P. ovale*).
- Lipophilic.
- Part of normal cutaneous flora.
- *Malassezia pachydermatis* causes similar infections.
- Well-recognized complication of total parenteral nutrition.
- Preterm and infants less than 12 months old.

Clinical manifestations

- Fever and/or apnoea and brachycardia.
- Interstitial pneumonia and thrombocytopenia.
- No sign of infection at catheter insertion sites.
- Predominant pathological changes involve heart and lungs.

Essential investigations

Microscopy and culture

Culture can be taken from blood that has passed through the catheter, with isolation from the catheter tip.

Subculture onto lipid-containing media, after which identifiable colonies can be seen after 4–6 days at 32°C.

Malassezia pachydermatis will often grow on lipid-free medium.

Microscopy reveals yeast cells with characteristic budding on a broad base.

Management

Remove the infected vascular catheter and discontinue the lipid supplements. Patients can be treated with parenteral fluconazole 5 mg/kg.

More stable patients can be given fluconazole or itraconazole by mouth.

Selected reading: recently published texts and monographs

General texts

Kibbler, C.C., Mackenzie, D.W.R., Odds, F.C. (eds) (1996) *Principles and Practice of Clinical Mycology*. John Wiley, Chichester.

Midgely, G., Clayton, Y.M., Hay, R.J. (1997) *Diagnosis in Colour: Medical Mycology*. Mosby-Wolfe, Chicago.

Richardson, M.D., Warnock, D.W. (1997) *Fungal Infection: Diagnosis and Management*, 2nd edn. Blackwell Science, Oxford.

Ajello, L. & Hay, R.J. (eds) (1998) *Topley & Wilson's Microbiology and Microbial Infections*, Vol. 4. *Medical Mycology*. Arnold, London.

Monographs on particular aspects

Elewski, B. (1998) *Cutaneous Fungal Infections*, 2nd edn. Blackwell Science, Oxford.

Roberts, D.T., Evans, E.G.V., Allen, B.R. (1998) *Fungal Infection of the Nail*, 2nd edn. Mosby-Wolfe, Chicago.

Richardson, M.D. & Elewski, B. (2000) *Fast Facts: Superficial Fungal Infections*. Health Press, Oxford.

Brakhage, A.A., Jahn, B., Schmidt, A. (eds) (1999) *Aspergillus fumigatus: Biology, Clinical Aspects and Molecular Approaches to Pathogenicity*. Karger, Basel.

Identification manuals

Campbell, C.K., Johnson, E.M., Philpot, C.M., Warnock, D.W. (1996) *Identification of Pathogenic Fungi*. Public Health Laboratory Service, London.

St-Germain, G. & Summerbell, R. (1996) *Identifying Filamentous Fungi. A Clinical Laboratory Handbook*. Star Publishing, Belmont.

Selected reading

Kane, J., Summerbell, R.C., Sigler, L., Krajen, S., Land, G. (1997) *Laboratory Handbook of Dermatophytes. A Clinical Guide and Laboratory Manual of Dermatophytes and Other Filamentous Fungi from Skin, Hair and Nails*. Star Publishing, Belmont.

De Hoog, G.S. & Guarro, J. (eds) (1999) *Atlas of Clinical Fungi*, 2nd edn. Centraalbureau voor Schimmelcultures, Baarn.

Sutton, D.A., Fothergill, A.W., Rinaldi, M.G. (1998) *Guide to Clinically Significant Fungi*. Williams and Wilkins, Baltimore.

WWW sites

Please note that this list is by no means exhaustive!

Fungal infections, general

http://www.mycology.adelaide.edu.au
http://medicine.bu.edu/fungal.htm
http://fungus.utmb.edu/mycology
http://www.keil.ukans.edu/~fungi
http://www.medsche.wisc.edu/medmicro/myco/mycology.html
http://www.topleyonline.com (Medical Mycology: Topley & Wilson's *Microbiology and Microbial Infections*, Volume 4 updates)

Specific infections

http://www.aspergillus.man.ac.uk
http://www.panix.com/~candida/
http://alces.med.umn.edu/Candida.html

Training

http://www.montana.edu/wwwmb/docs/Medicalmycology.html

Societies

http://www.asmusa.org/division/f/divf_main.htm (American Society for Microbiology, Division F: Medical Mycology)
http://www.leeds.ac.uk/isham/ (International Society for Human and Animal Mycology)

Publishers

http://www.blacksci.co.uk (medical mycology books and journals)
http://www.blacksci.co.uk/products/journals/jmvm.htm (Medical Mycology. The journal of the International Society for Human and Animal Mycology)

http://www.prous.es/product/journal/mic.html (Current Topics in Medical Mycology)

http://www.RevIberoamMicol.com (ejournal: Revista Ibero-america de Micologia)

Index